MATHEMATICS
The
Language Concepts

MATHEMATICS
The
Language Concepts

by

John A. H. Anderson, BA PhD

UNIVERSITY OF BRADFORD

* * *

Stanley Thornes (Publishers) Ltd.

First published in 1974 by Stanley Thornes (Publishers) Ltd
17 Quick Street London N1 8HL

ISBN 0 85950 008 X   (cased edition)
ISBN 0 85950 009 8   (paperback edition)

Text set in 10/12 pt. Monotype Times New Roman, printed by photolithography,
and bound in Great Britain at The Pitman Press, Bath

INTRODUCTION

This book is designed for students who are required to produce mathematical proofs, for use during the beginning of their undergraduate career. It may profitably be read during the vacation before attending University.

The objective of the book is to set up an attitude to mathematics appropriate for a University course, and to explain in detail certain preliminaries essential to the pursuit of modern mathematics.

The student is advised not to strive for detailed comprehension of Chapter 1 on the first reading. This chapter is intended to be read both before and after the other chapters.

CONTENTS

Introduction   v

Contents   vii

CHAPTER 1

The Nature of Mathematical Models

1. The Problem of Reality   1
2. Prediction Using Models   2
3. The Description of a Mathematical Model   5

CHAPTER 2

Statements

1. A Primitive Analysis of Language   9
2. Truth Tables   12
3. Tautologies   18
4. A Broad View of Connectives   19

CHAPTER 3

A Description of Proof

1. Rules of Inference   23
2. Direct Proof   27
3. Indirect Proof   28
4. Proof by Contradiction   30
5. Use of Equivalent Statements   32

CHAPTER 4

Predicates

1. Variables and Parameters   36
2. Defining Sets   40
3. Compounding Predicates and Sets   42

CHAPTER 5

Quantification

1. Quantifiers   52
2. Proof of Quantified Statements   58

3. The Induction Theorem 64
4. Set Equality and Inclusion 71
5. The Power Set 76

CHAPTER 6

Relationships

1. Relationships as Objects 77
2. Properties which Characterise Relations 87
3. Equivalence Relations 95
4. Order Relations 98
5. Functional Relations 104

CHAPTER 7

The Axiomatic Approach

1. The Language of Quantification 112
2. The Language of Sets 117

CHAPTER 1

THE NATURE OF MATHEMATICAL MODELS

§1 The Problem of Reality

We may examine the real world only with our senses. The instruments which we use as extensions of those senses are developed only after a certain understanding of the nature of reality has been achieved. The simplest assumption about this real world is that it contains entities and that these entities behave according to certain rules. The initial objective of science was to find out what these entities were and to discover the rules which governed their behaviour. However, certain difficulties arise even before we begin the search.

Firstly, entities may exist which we cannot perceive directly but which nevertheless affect the universe in which we live and thus are apparent to us through their effects. Everyone has sensed effects which have been explained in terms of an entity called an electron. Yet no-one has seen an electron, no-one can testify that such a thing exists. Rather it is part of a model we have invented to explain certain effects, and it takes different forms in different models. Even though we may use a model to construct an instrument to "count electrons", our interpretation of what the instrument is doing depends on our acceptance of the original model. We cannot construe the results as proving the existence of electrons. We cannot hope to describe the entities which give rise to such effects in terms of what they look, smell, taste, sound or feel like, while our senses will not perceive them. Neither would we expect to be able to describe how these entities achieve their effects.

Secondly, although we may observe a phenomenon, how can we be sure that we are not deceived? Our colour vision is based on being able to detect three colours. The other colours we perceive are our brain's interpretation of combinations of these three primary colours. Certain combinations are interpreted as colours which do not occur in the visible spectrum. How real are these colours? Again, we may obtain different impressions of a piece of terrain by obtaining an aerial photograph, an aerial photograph using film sensitive to infra red light and a radar image. The data contained in the first photograph will be readily interpreted by the brain, while the infra red photograph and the radar image will be easily interpreted only by a practical operator. Again, the image thrown on the retina by the eye's lens is upside down yet our interpretation of this information yields an erect image. In a famous experiment, a man was fitted with goggles which ensured that the image thrown on his retina was erect. As a result, he saw everything upside down until after a time his brain learned to reinterpret the information presented to it and provided him with an erect image. There are examples of individuals whose inability to read appears to stem from their brain's refusal to settle upon a fixed interpretation of the visual data it receives. Such a person may sometimes reach for the right hand side of a door although the handle is on the left, while at other times finding the handle successfully.

It would seem that a quest for absolute truth is doomed to disappointment, but perhaps we should look for more easily defined goals and ask what the purpose of science is. One point of view is that the object of science is to control our environment, for the purposes of comfort, convenience, amusement or survival. Evidently we shall perceive effects which are relevant to these purposes and the distinction between real and deceived perception becomes a question of interpretation. Since this is not intended to develop into either a philosophical or medical discussion, let us restate our objectives in a very practical way. Firstly, given that we can perceive a certain situation at a point in time, we should like to be able to predict how we shall perceive the situation as it develops in time. Secondly, given a desirable situation of which we have conceived, we should like to know how to arrange matters so that this desirable situation results. Also we should like to know whether this desirable situation develops into other desirable situations or whether it is a fool's paradise.

§2 Prediction Using Models

In the laboratory, we put ourselves in situations where we are likely to observe things, but the vast majority of detail goes undetected. Much of it indeed may be undetectable and we can never know what is happening. We have however a method for predicting the observable portion. We construct a mathematical model of the real situation. The objects in this model are built up out of numbers and collections of numbers. Thus a three dimensional vector (1, 3, 5) or a 2×2 matrix $\begin{pmatrix} 1 & 2 \\ 0 & 5 \end{pmatrix}$ are mathematical objects. The behaviour of the objects in the model is governed by rules of our own making. Such a model is constructed so that it behaves in the same way as the real situation behaves, in so far as we can observe the behaviour of the real situation.

One might be tempted to believe that the behaviour of the model ought to be completely known to us, since it is our own creation. In fact, although the rules governing the behaviour of the model are laid down by us, their consequences may not be obvious.

Some models are created which do not behave like any real situation; they are created for the satisfaction of their creators and not to mirror a real situation. However, since their creators live in the real world and draw much inspiration from it, it is not surprising that a model created for the satisfaction of its creator may later be found to model a real situation.

When we are motivated by the real world to create a mathematical model, we usually have a good idea about which observable real entities correspond to which mathematical entities. Most often however there are many mathematical entities involved, which form a substratum of the model in the sense that they do not correspond to any observable real entities. These entities are introduced into the model, not to correspond to a real object, but to make the model work. It is tempting to ask what sort of real objects these mathematical entities might correspond to, and to try to imagine suitable real objects. This can help one's intuitive idea of how the mathematical objects ought to behave. For example, we may build up a model of a gas in which the gas is made up of three dimensional time dependent vectors to each of which is attached a positive number (mass). It is helpful to visualize these vectors as small billiard balls moving about, but the vectors may not correspond to real objects at all. We are content if observable phenomena like the lengths of certain columns of mercury (temperature and pressure) can be accurately predicted.

We have considered a vertical division between the real world and a model of the real world and also a horizontal division which splits the real world into observed and unobserved parts and splits the model into those parts which correspond to the observable real world and those which do not.

The process of predicting the behaviour of the real world takes the following form. First we choose a suitable model. Next we translate the observed real situation whose development we wish to predict into the corresponding situation in the model. We then develop the situation in the model. The features of the developed mathematical situation which correspond to observable reality are then translated back to the real world and the result constitutes our prediction of the development of the observable real situation. This process is problematical. Firstly, we have to put ourselves in a position to make observations of the real situation in a systematic manner, and be able to describe these observations with precision. Secondly, we have to define a suitable model. For this we have to spot patterns in our observations in order to ensure that the reality which corresponds to the model accords with our observations, and to test some of the minor predictions of the model. Thirdly, we have to be able to develop the situation in the model. Almost always we only develop the situation approximately. We may even be forced to adopt a model because it is simple to develop even though we know of one which fits the real situation better. Thus we continue to use Newtonian mechanics for certain real situations even though the relativistic model is a better fit. It is sometimes useful to develop a well fitting though complicated model from a qualitative point of view, and obtain quantitative information from a simpler model. Fourthly and lastly we must decide what features of the model we may expect to correspond to observable reality, bearing in mind that we shall have to add features of the model's substratum in translating from observational reality to the model, and subtract such features in translating from the model back to observational reality. We offer an illustration.

A steel ball is dropped from a certain height onto a floor of solid steel. We may imagine a point and a line in Euclidean space and supply the usual framework of Newtonian mechanics and elasticity. There is a good deal more involved in producing this model but we shall rely on the reader's intuition. The model predicts that the point will bounce on the line, the height of each bounce being a fixed fraction of the height of the preceding bounce. Thus the model predicts an infinite number of bounces although the time taken for all the bounces will be finite. It is clear that although the point in the model bounces an infinite number of times, all but the first few will involve so little movement and so little time that even were the real ball to execute these bounces they would be unobservable. The model is successful if it predicts the observable bounces correctly.

Errors of various kinds affect the accuracy of our predictions. Observational errors arise both in the initial observations and while testing the model by comparing its predictions with reality. Almost always we shall have calculating errors due to the difficulty of developing the model, which difficulty often necessitates approximation. Error may be inherent in the model due to an unsatisfactory representation of the objects in the model. The representation of real numbers as infinite decimals is an example, since we usually have to approximate irrational numbers by rational numbers in calculation. Finally, the model itself may be a poor one in that it does not model reality very closely.

At this point we stress that a mathematical model is a complete creation in its own right. There is no direct connection between such a model and the reality it purports to model. The connection exists only in the mind of the scientist using the model. There is no convincing reason of which the author is aware why we should be able to connect reality with a mathematical model, neither can we be sure that a model which fits today will fit tomorrow. Nevertheless, with all its difficulties, the modelling process provides us with a surprisingly satisfactory method of prediction over a wide range of phenomena.

§3 The Description of a Mathematical Model

A description of the real world is often vague. One reason why this is inevitable is that we simply do not know enough to give a full description. Another is that even a description of what we do know is liable to be lengthy if it is precise. Consequently such descriptions depend greatly on the listener's experience and intuition to fill in the gaps which the description leaves. A description of a mathematical model however must be precise and complete, since the model is just what we say it is and no more. The listener cannot fill in gaps since the model is completely the invention of the describer. If the describer omits to mention a relationship, then that relationship does not exist in the model. If the description is vague, then we will not be able to determine how the model works. A description of the real world describes things which already exist and which the listener may inspect. A description of a mathematical model on the other hand defines the model.

One can refer to a real object by pointing to it. If one wishes to refer to an object often, it is usual to give the object a name. This obviates the necessity of having the object present. It is characteristic of a real object that there is only one of it; copies differ from one-another at least in not being in the same place at the same time. Thus giving a name to an object enables several people in different places to refer to it. The advantage of a name is that it is readily reproducible and loses nothing in the reproduction.

In contrast, a mathematical object has no "real" existence. One cannot point to it. Its description must therefore be complete, and it can only be referred to via its description or the name given to, it. It follows that the language used to describe mathematical objects must be capable of precision. Just as a description of a real object will fail if the two people involved have different sensory equipment, so a description of a mathematical object will fail if they do not interpret the language in the same way.

It is apparent that the naming of mathematical objects is very important, since all we ever see of these objects are their names. Of course we give objects nick-names for brevity, but the full name of a mathematical object should contain all the essential features of the object. This is necessary so that the relationships between the objects can be described in terms of the full name of the objects. The function of a playing card is to carry its name. Card games may be played with any 52 different objects, without so much as altering the descriptions of the rules of the games, simply by assigning the appropriate 52 names to the new objects.

Our language must also be capable of describing precisely the rules which the mathematical objects have to obey. We shall introduce a vocabulary together with the method by which the words are strung together to form sentences. We can form sentences which describe behaviour only, in the sense that although they are used to describe relationships between objects, they do not actually mention any specific objects nor any specific relationships. Given a list of such sentences, our problem is to find a collection of mathematical objects and relationships such that the sentences are true when regarded as sentences about these objects and relations. Thus we have a description first and only afterwards do we look for a model which fits the description. The description itself may be formulated as a description of the real world, but since no specific objects or relationships are mentioned we can ask whether there exists a mathematical model which fits the description, and indeed how many models fit the description. If there is no model, our description was contradictory. If there are many models we might take the view that our description is incomplete, and that we need to extend it in order to discard some models on the grounds that they do not fit the extended description. On the other hand, it may be that our description covered a real observable situation and that the models differ only in their substrata.

This would mean that for the purpose of predicting observable reality, any of these models would do. This gives us some freedom of choice of model to use at our convenience. It may be thought that in the ideal situation exactly one model fits the description. In such a case we call the description *categorical*. We must however be careful, for, as we now explain, one can always produce multitudes of different models to fit any description which is not contradictory. What we mean when we say that there is only one model which fits a categorical description is that all the models which fit this description are essentially the same in the following sense.

There are many card games, all defined by lists of rules. We use cards to play these games because they are easy to handle and besides the rules of the games are all described in terms of the suits and numbers printed on the cards. It would however be possible to play essentially the same games with any fifty-two different objects. Many people have been driven by boredom and a lack of packs of cards into doing exactly this. There are two methods of dealing with the new objects. One might change the form of the rules so that instead of being explained in terms of playing cards they are explained in terms of the new objects. This procedure would be very complicated should we be attempting to play bridge with fifty-two different types of chair. The alternative method is to change the names of the new objects to those of the old, leaving the rules alone. Thus we might label a brown armchair as the three of diamonds.

We hope it is now apparent that it does not matter what mathematical objects we populate our model with, it remains essentially the same model. It is the relationships between the objects which are important. Another explanation leads to the same conclusion. A model of something is supposed to resemble it in some way. A real object is known to us through our senses. The mathematical object in the model which corresponds to the real object does not look, taste, smell, feel or sound like the real object. How then can it resemble it? The answer is that it behaves like it. The relationships which the mathematical object has with the other mathematical objects in the model mirror the relationships which the real object has with other real object. There is one obvious question left, namely why are there so many different kinds of mathematical objects and what determines the kind of mathematical object we choose when we assemble a model? The kind of object we choose depends on the kind of relationships between the objects which we wish to describe. Given the relationships, we choose objects so that we can easily describe the relationships in terms of the objects, in the same way as we have named the fifty-two playing cards in such a way that we can easily describe the rules of a card game in terms of the names of the playing cards.

CHAPTER 2

STATEMENTS

§1 A Primitive Analysis of Language

DEFINITION 1. A *statement* is a sentence in everyday language which is either true or false but not both. Thus sentences which are questions are not statements.

We shall use small letters p, q, r, ... to denote statements and we shall refer to them as *elementary statements*.

From the elementary statements we may make up more complicated statements using the words "not", "and", "or" and "implies". Thus from the two elementary statements "p" and "q" we may make the statements "not p", "p and q", "p or q" and "p implies q". We may construct progressively more complicated statements such as "not (p and q)", "(p and q) implies (p or q)" and so on. The manufactured statements are called *compound statements*, to distinguish them from elementary statements. We shall analyse the compound statements to see how the truth and falsity of a compound statement depends on the truth and falsity of the elementary statements from which it was obtained. We do not discuss the nature of the elementary statements p, q, r ... neither are we interested in whether these statements are actually true or false. Rather, we shall consider all the possible combinations of truth and falsity of the elementary statements and examine what effect each combination has on the compound statements. For example, the compound statement "p or (not p)" is true whether p is true or false.

We shall introduce symbols for the four words "not", "and", "or" and "implies". These symbols are not mere shorthand. Each of these four words is capable of various shades of meaning as they are used in everyday language, but our symbols will have a unique meaning, the meaning appropriate to mathematical proof, and that meaning we shall precisely explain.

The symbols are $\sim$ for "not", $\wedge$ for "and", $\vee$ for "or" and $\Rightarrow$ for "implies".

Let p denote the statement "2 is an even number". The compound statement "$\sim p$" could be rendered as "2 is not an even number", "it is false that 2 is an even number", or "it is not true that 2 is an even number". Our symbolism gives us a form for these statements which does not involve our knowing the nature of p.

The statement "$p \Rightarrow q$" which we read "p implies q" may also be rendered "If p then q", "p only if q", "q if p", "q because p", or "q follows from p". These different forms commonly carry different shades of meaning which are not relevant to the one in which we are interested. From a mathematical point of view, they are equivalent to our "$p \Rightarrow q$".

The symbols $\sim$, $\wedge$, $\vee$, $\Rightarrow$ are called *connectives*. The symbol $\sim$ we shall call a *unary connective* since it requires just one statement "p" in order to produce a compound statement "$\sim p$". The symbol $\wedge$, $\vee$ and $\Rightarrow$ we call *binary* connectives since each requires two statements, or the same statement twice, in order to produce a third statement.

We have a process whereby from elementary statements p, q, $r \ldots$ we can generate as many new statements as we please. For example, $p \wedge q, p \Rightarrow q, p \vee (p \wedge q), \sim(p \Rightarrow q), p \wedge (p \Rightarrow q)$. Here are our compound statements in symbols. Notice the use of brackets to avoid ambiguity. The statement $p \wedge (p \Rightarrow q)$ is different from the statement $(p \wedge p) \Rightarrow q$ with the result that to write $p \wedge p \Rightarrow q$ would be ambiguous.

We have now a method of providing a crude analysis of mathematical statements. We are analysing the logical structure of these statements. An example of such analysis follows.

Consider the statement "3 divides the product 171×227 only if 3 divides 171 or 3 divides 227". Let p be "3 divides the product 171×227", q be "3 divides 171" and r be "3 divides 227".

We can now render the statement as "p only if $q \vee r$" and the complete analysis is "$p \Rightarrow (q \vee r)$". Notice that we have had to provide the brackets, which were somehow understood in the initial form of the statement. The purpose of such an analysis is to aid proof, as we shall later describe.

Exercises 1

Analyse the following statements.

1. If 3 is odd and 4 is even then 3×4 is even.
2. If 13 is a prime number and 4 divides 13 then $4 = 1$ or $4 = 13$.
3. If 4 is not an even number then 2 does not divide 4.

4. The number 7 is odd if and only if 7^2 is odd.
5. If 3 divides 171×227 and 3 does not divide 227 then 3 divides 171.

Answers 1

1. Let p be "3 is odd", q be "4 is even" and r be "3×4 is even". The analysis is $(p \wedge q) \Rightarrow r$.
2. Let p be "13 is a prime number", q be "4 divides 13", r be "$4 = 1$" and s be "$4 = 13$". The analysis is $(p \wedge q) \Rightarrow (r \vee s)$.
3. Let p be "4 is an even number", and q be "2 divides 4". The analysis is $(\sim p) \Rightarrow (\sim q)$.
4. Let p be "The number 7 is odd" and q be "7^2 is odd". The analysis is $(p \Rightarrow q) \wedge (q \Rightarrow p)$.
5. Let p be "3 divides 171×227", q be "3 divides 227" and r be "3 divides 171". The analysis is $(p \wedge (\sim q)) \Rightarrow r$.

Had we written the last analysis as $p \wedge (\sim q) \Rightarrow r$ or even $p \wedge q \Rightarrow r$ the reader would probably have understood what was intended. We may adopt conventions in order to simplify bracketing. For example, in the absence of brackets, $\sim$ is taken to apply to the letter immediately succeeding it. Thus $\sim p \wedge q$ means $(\sim p) \wedge q$ rather than $\sim(p \wedge q)$. Again, if $\Rightarrow$ occurs only once in a statement such as $p \wedge \sim q \Rightarrow r \vee s$ then we take the statement to mean $(p \wedge \sim q) \Rightarrow (r \vee s)$ rather than, for example, $(p \wedge (\sim q \Rightarrow r)) \vee s$, or $((p \wedge \sim q) \Rightarrow r) \vee s$.

In answer to the third question one might have given an analysis as follows. Let p be "4 is not an even number" and q be "2 does not divide 4". The analysis would then have been "$p \Rightarrow q$". This is quite proper, though less useful than the more detailed analysis given. The choice of which statements are to be regarded as elementary is up to the analyst, but clearly the more structure he can expose the better the analysis. We take the opportunity to introduce some more notation. Whereas $p, q, r, \ldots$ denote elementary statements, we use $P, Q, R, \ldots$ to denote statements which may be elementary or may be compounded from the elementary statements. Thus we can use P for a compound statement when we do not wish to expose the structure of P in terms of elementary statements and connectives. We can use this notation immediately to describe exactly what statements can be produced from the elementary statements.

DEFINITION 2.   (i) $p, q, r, \ldots$ are statements.
             (ii) Whenever P and Q are statements then $(\sim P)$, $(P \wedge Q)$, $(P \vee Q)$ and $(P \Rightarrow Q)$ are statements.

Thus we have an algebraic description of statements, which does not mention truth and falsehood.

§2 Truth Tables

We label each elementary statement 1 or 0. In an argument, the labelling of a statement with 1 corresponds to assuming that the statement is true while labelling it 0 corresponds to assuming it false. We describe the label of a statement P as its *truth value*. For a given labelling of the elementary statements we wish to define a method for deducing the label appropriate to each of the compound statements which can be made up from those elementary statements and the connectives. Since the compound statements are made using $\sim$, $\wedge$, $\vee$ and $\Rightarrow$ we shall show how, given the labelling of two statements P and Q, we can deduce the label appropriate to $\sim(P)$, $(P \wedge Q)$, $(P \vee Q)$ and $(P \Rightarrow Q)$.

Negation. From a knowledge of the label of P we can deduce the label of $\sim(P)$ from the following table, called a *truth table*.

	P	$(\sim P)$
First line	0	1
Second line	1	0

The first line states that if P is labelled 0 then $(\sim P)$ is labelled 1, i.e. If P is false then $(\sim P)$ is true. The second line states that if P is true then $(\sim P)$ is false.

Conjunction. Given a labelling of P and Q we can deduce the label of $(P \wedge Q)$ from the following truth table.

P	Q	$(P \wedge Q)$
0	0	0
0	1	0
1	0	0
1	1	1

Thus $(P \wedge Q)$ will be true only if both P and Q are true. Notice that the left hand columns cover all possible labellings of P and Q and do so in a systematic manner, for 00, 01, 10, and 11 are in fact the binary representations for 0, 1, 2, 3. By always using this standard form for the left hand columns, we can speak of the right hand column as "*the* truth table for $(P \wedge Q)$". A proof of a statement of the form $(P \wedge Q)$ will normally consist of a proof of P together with a proof of Q.

Disjunction. Given a labelling of P and Q we can determine the label of the disjunction $(P \vee Q)$ using the following table.

P	Q	$(P \vee Q)$
0	0	0
0	1	1
1	0	1
1	1	1

Thus $(P \vee Q)$ is false only if both P and Q are false. The previous two tables give interpretations of "not" and "and" which agree very closely with the normal usage of these words. A frequent use of "or" occurs in the statement "You can go to the pictures or you can go swimming". This statement carries the impression that you can't do both. This impression is not conveyed by our meaning of $\vee$, as the last line of the table shows. The meaning we have chosen for $\vee$ is that most often needed in mathematics. The other meaning is conveyed by the more complicated statement $(P \vee Q) \wedge \sim(P \wedge Q)$, "$P$ or Q and not both P and Q". We can deduce the truth table of this statement from the truth tables of $\sim$, $\wedge$ and $\vee$ as follows.

EXAMPLE 1

P	Q	$(P \vee Q)$	$(P \wedge Q)$	$\sim(P \wedge Q)$	$(P \vee Q) \wedge \sim(P \wedge Q)$
0	0	0	0	1	0
0	1	1	0	1	1
1	0	1	0	1	1
1	1	1	1	0	0

The columns for $(P \vee Q)$ and $(P \wedge Q)$ are deduced directly from the columns for P and Q using the tables for $\vee$ and $\wedge$. The column for $\sim(P \wedge Q)$ is obtained from that for $(P \wedge Q)$ using the table for $\sim$. Finally, the column for $(P \vee Q) \wedge \sim(P \wedge Q)$ is obtained from the columns for $(P \vee Q)$ and $\sim(P \wedge Q)$ using the table for $\wedge$.

A proof of a statement $(P \vee Q)$ could in fact consist of either a proof of P or else a proof of Q. This would be a rather facile situation, since if the proof is a proof of P, why call it a proof of $(P \vee Q)$? Normally a proof of $(P \vee Q)$ will show that either P or Q is true but will fail to specify which one is true. It may in fact show that if P is false then Q must be true. Thus it will be a proof of $\sim P \Rightarrow Q$.

Implication. The idea of implication lies at the root of the argument. Whether an argument is sound or not will depend on the view taken of implication. Given the labels of P and Q, the label of $(P \Rightarrow Q)$ is determined by the following table.

	P	Q	$(P \Rightarrow Q)$
Line 1	0	0	1
Line 2	0	1	1
Line 3	1	0	0
Line 4	1	1	1

Thus the implication $(P \Rightarrow Q)$ is false only when P is true and Q is false. Many people find that this view of implication does not accord with their intuitive view. Intuition is based on experience and one cannot talk people out of their experiences. One can however add to their experiences.

When providing a proof of $(P \Rightarrow Q)$ we do not know the truth values of P and Q. If we knew that P was true, then what we would provide is a proof of Q rather than a proof of $(P \Rightarrow Q)$. If we knew P was false, there would seem little point in proving $(P \Rightarrow Q)$. This last remark applies also if we know the truth value of Q. Thus our experience of proving $(P \Rightarrow Q)$ is limited to the situation when the truth values of P and Q are unknown. Thus if we are to prove $(P \Rightarrow Q)$ then we must prove that $(P \Rightarrow Q)$ is true, no matter what the truth values of P and Q may be.

The most popular method of proving $(P \Rightarrow Q)$ is called the *direct* method. This is a demonstration that if P is true then Q must be true. Having accomplished this, we are content to declare $(P \Rightarrow Q)$ true. But have we really finished? We have not considered the possibility that P may be false. Since we are ignorant of the truth value of P, this possibility is very real yet we have ignored it. That the possibility of P being false is ignored is a consequence of lines 1 and 2 of our table for $\Rightarrow$. For this assures us that if P is false, then $(P \Rightarrow Q)$ is true and there is nothing left to prove. This also explains why although any mathematician has had the experience of arguing $(P \Rightarrow Q)$ from the premise that P is true, none have had the experience of arguing $(P \Rightarrow Q)$ from the premise that P is false.

If the key to proving $(P \Rightarrow Q)$ lies in showing that we cannot have P true and Q false, two other methods of proving $(P \Rightarrow Q)$ become apparent. In the direct method, we assigned the label 1 to P and showed that then Q cannot have the label 0. Symmetrically, we can assign the label 0 to Q and prove that P cannot then have the label 1. This is called the *indirect* method of proving $(P \Rightarrow Q)$. It begins by assuming that Q is false and deduces that P must be false.

The last method of proving $(P \Rightarrow Q)$ is called *proof by contradiction*. This attacks line 3 of the table directly. Under the assumption that P is labelled 1 and Q labelled 0 we deduce that a statement R has both the label 0 and the label 1. This is not seemly behaviour for a statement according to the first definition of §1. Hence we cannot label P with 1 and Q with 0. This proves $(P \Rightarrow Q)$ since any other labelling of P and Q assigns the label 1 to $(P \Rightarrow Q)$.

Our experience of executing these proofs leaves us with impression that if $(P \Rightarrow Q)$ is true then there must be some connection between P and Q. If we are to prove $(P \Rightarrow Q)$ in ignorance of the truth values of P and Q, it is difficult to see how there may not be a connection. However, if we can prove that P is false then we have immediately a proof that $(P \Rightarrow Q)$; alternatively a proof that Q is true provides us immediately with a proof of $(P \Rightarrow Q)$. In neither case need there be any connection between P and Q. Of course we have no experience of such methods of proving $(P \Rightarrow Q)$ since there is no apparent value in proving $(P \Rightarrow Q)$ when we know either that P is false or that Q is true.

Let p be the statement "$1 = 2$" and q the statement "$3 = 3$". We wish to prove $(p \Rightarrow q)$. Assuming that we do not know the truth values of p and q we proceed as follows. Let p be true. Then $1 = 2$. Therefore $2 = 1$, since changing the order of a true equation always produces a true equation. We may produce a true equation from two other true equations by equating the sum of their left hand sides to the sum of their right hand sides. Hence $3 = 3$ is true. Here we have blinded ourselves to the truth values of p and q and thereby forced ourselves to provide a chain of reasoning, or connection, between p and q in order to prove $(p \Rightarrow q)$. This chain would have been trivial had we admitted that we knew q to be true for after assuming p to be true we could have immediately asserted the truth of q thus completing the chain. We examine chains of reasoning later in the section on rules of inference.

EXAMPLE 2.   Find the truth table of
$$(p \Rightarrow q) \wedge (q \Rightarrow p).$$

p	q	$(p \Rightarrow q)$	$(q \Rightarrow p)$	$(p \Rightarrow q) \wedge (q \Rightarrow p)$
0	0	1	1	1
0	1	1	0	0
1	0	0	1	0
1	1	1	1	1

Notice that in obtaining the column for $(q \Rightarrow p)$, care must be taken in using the table for $\Rightarrow$ that the values of q and p are used in the correct order. The statement $(p \Rightarrow q) \wedge (q \Rightarrow p)$ is often abbreviated to $(p \Leftrightarrow q)$, read "p is equivalent to q". We may regard $\Leftrightarrow$ as a connective whose truth table is given by

p	q	$(p \Leftrightarrow q)$
0	0	1
0	1	0
1	0	0
1	1	1

Thus $(p \Leftrightarrow q)$ is true only when p and q have the same truth values.

EXAMPLE 3.   Find the truth table for
$$((p \Rightarrow q) \wedge (q \Rightarrow r)) \Rightarrow (p \Rightarrow r).$$
Denote the statement $(p \Rightarrow q) \wedge (q \Rightarrow r)$ by S.

p	q	r	$(p \Rightarrow q)$	$(q \Rightarrow r)$	S	$(p \Rightarrow r)$	$S \Rightarrow (q \Rightarrow r)$
0	0	0	1	1	1	1	1
0	0	1	1	1	1	1	1
0	1	0	1	0	0	1	1
0	1	1	1	1	1	1	1
1	0	0	0	1	0	0	1
1	0	1	0	1	0	1	1
1	1	0	1	0	0	0	1
1	1	1	1	1	1	1	1

The triples 000, 001, 010 and so forth are in fact numbers zero to seven in binary form. Notice that this statement has truth value 1, no matter what truth values are assigned to p, q and r.

Exercises 2

1. Obtain the truth tables for the following statements.
   (i) $((p \wedge q) \wedge r) \Leftrightarrow (p \wedge (q \wedge r))$.
   (ii) $(p \wedge (p \Rightarrow q)) \Rightarrow q$.
   (iii) $(p \Rightarrow q) \Leftrightarrow (\sim p \vee q)$.

2. Given that the compound statement $(\sim(p \Rightarrow q)) \wedge r$ has truth value 1, what must the truth values of p, q and r be?

3. Given that the statements $(p \wedge q)$, $(q \Rightarrow r)$ and $(r \wedge p) \Rightarrow s$ are all true, what is the truth value of s?

4. For what truth values of p and q is the implication $(p \vee q) \Rightarrow \sim (p \wedge q)$ false?

Answers 2

1. (i)

p	q	r	$(p \wedge q)$	$(p \wedge q) \wedge r$	$(q \wedge r)$	$p \wedge (q \wedge r)$	$\Leftrightarrow$
0	0	0	0	0	0	0	1
0	0	1	0	0	0	0	1
0	1	0	0	0	0	0	1
0	1	1	0	0	1	0	1
1	0	0	0	0	0	0	1
1	0	1	0	0	0	0	1
1	1	0	1	0	0	0	1
1	1	1	1	1	1	1	1

Notice that we have used the truth table for $\Leftrightarrow$, also that $(P \Leftrightarrow Q)$ will be true for all truth values of the elementary statements in P and Q provided that P and Q have the same truth table.

(ii)

p	q	$(p \Rightarrow q)$	$p \wedge (p \Rightarrow q)$	$(p \wedge (p \Rightarrow q)) \Rightarrow q$
0	0	1	0	1
0	1	1	0	1
1	0	0	0	1
1	1	1	1	1

(iii)

p	q	$p \Rightarrow q$	$\sim p$	$\sim p \vee q$	$\Leftrightarrow$
0	0	1	1	1	1
0	1	1	1	1	1
1	0	0	0	0	1
1	1	1	0	1	1

The truth tables of $(p \Rightarrow q)$ and $(\sim p \vee q)$ are the same.

2. If $(\sim(p \Rightarrow q)) \wedge r$ is true then both $\sim(p \Rightarrow q)$ and r are true. If $\sim(p \Rightarrow q)$ is true then $(p \Rightarrow q)$ is false. From the truth table for $\Rightarrow$, this occurs only when p is true and q is false. Thus p, q and r must have truth values 1, 0 and 1 respectively.

3. Firstly, $(p \wedge q)$ is true so p and q must each be true. Now q and $(q \Rightarrow r)$ are true. From the truth table for $\Rightarrow$ we see that this situation occurs only when r is true. Since r and p are true, so is $(r \wedge p)$. But $(r \wedge p) \Rightarrow s$ is given true. Hence s is true. Thus s has truth value 1.

4. If $(p \vee q) \Rightarrow \sim(p \wedge q)$ is false then from the truth table for $\Rightarrow$, $p \vee q$ must be true while $\sim(p \wedge q)$ must be false. Hence $(p \wedge q)$ must be true. Therefore p and q must each be true. Thus the implication is false when p and q each have truth value 1.

§3 Tautologies

DEFINITION 3. A *tautology* is a compound statement whose truth value is 1, no matter what truth values the elementary statements within it are given.

Among the statements which we can produce using our connectives $\sim$, $\wedge$, $\vee$ and $\Rightarrow$, tautologies are the most important type of statement. One reason for this is mentioned in the section on rules on inference. Among the tautologies we find a particularly interesting type of tautology, namely those of the form $(P \Leftrightarrow Q)$. These give us a definition of equivalence or equality between statements.

DEFINITION 4. Two statements P and Q are *equivalent* if and only if $(P \Leftrightarrow Q)$ is a tautology.

We are now in a position to say some interesting things about our connectives. We can regard the following equivalences as exhibiting algebraic properties of the connectives in them. The following are all tautologies.

   (i) $\sim(\sim p) \Leftrightarrow p$
  (ii) $p \vee q \Leftrightarrow q \vee p$
 (iii) $p \wedge q \Leftrightarrow q \wedge p$
 (iv) $(p \vee q) \vee r \Leftrightarrow p \vee (q \vee r)$
  (v) $(p \wedge q) \wedge r \Leftrightarrow p \wedge (q \wedge r)$
 (vi) $((p \Rightarrow q) \wedge (q \Rightarrow r)) \Rightarrow (p \Rightarrow r)$
(vii) $p \wedge (q \vee r) \Leftrightarrow (p \wedge q) \vee (p \wedge r)$
(viii) $p \vee (q \wedge r) \Leftrightarrow (p \vee q) \wedge (p \vee r)$.

Exercise 3

Prove that (vii) and (viii) above are tautologies.

Answer 3

(vii)

p	q	r	$(q \vee r)$	$p \wedge (q \vee r)$	$(p \wedge q)$	$(p \wedge r)$	$(p \wedge q \vee (p \wedge r))$
0	0	0	0	0	0	0	0
0	0	1	1	0	0	0	0
0	1	0	1	0	0	0	0
0	1	1	1	0	0	0	0
1	0	0	0	0	0	0	0
1	0	1	1	1	0	1	1
1	1	0	1	1	1	0	1
1	1	1	1	1	1	1	1

(viii)

p	q	r	$(q \wedge r)$	$(p \vee (q \wedge r))$	$(p \vee q)$	$(p \vee r)$	$(p \vee q) \wedge (p \vee r)$
0	0	0	0	0	0	0	0
0	0	1	0	0	0	1	0
0	1	0	0	0	1	0	0
0	1	1	1	1	1	1	1
1	0	0	0	1	1	1	1
1	0	1	0	1	1	1	1
1	1	0	0	1	1	1	1
1	1	1	1	1	1	1	1

§4 A Broad View of Connectives

We defined the behaviour of the connectives $\sim$, $\wedge$, $\vee$ and $\Rightarrow$ using truth tables. We defined the behaviour of the connective $\Leftrightarrow$ by saying that $(p \Leftrightarrow q)$ has the same truth table as $(p \Rightarrow q) \wedge (q \Rightarrow p)$. We might describe this last method of defining $\Leftrightarrow$ by saying that $\Leftrightarrow$ can be defined in terms of $\Rightarrow$ and $\wedge$. We wish to make two observations. First, some connectives can be defined in terms of others. Secondly, every connective is associated with the truth table which defines its behaviour.

The first question we can answer is, how many connectives are there? There are as many connectives as there are truth tables. Thus there are just $4 = 2^2$ unary connectives.

p	0	1	2	3
0	0	1	0	1
1	0	0	1	1

Number 1 is just unary connective $\sim$, i.e. $\sim p$ has the truth table of 1. Number 0 is equivalent to $p \wedge \sim p$, 2 is equivalent to p and 3 is equivalent to $p \vee \sim p$. Thus these four connectives can be defined in terms of $\sim$, $\wedge$ and $\vee$.

There are $16 = 2^4$ binary connectives.

p	q	0	1	2	3	4	5	...	14	15
0	0	0	1	0	1	0	1		0	1
0	1	0	0	1	1	0	0		1	1
1	0	0	0	0	0	1	1		1	1
1	1	0	0	0	0	0	0		1	1

The truth table for binary connective number 5 is obtained by writing 5 as a four figure binary number, $5 = 0.2^3 + 1.2^2 + 0.2^1 + 1.2^0 = 0101$, and arranging the zero's and one's in a vertical column with the right hand of 0101 at the top, thus

$$
\begin{array}{c}
5 \\
\hline
1 \\
0 \\
1 \\
0 \\
\end{array}
$$

The binary for 7 is $0.2^3 + 1.2^2 + 1.2^1 + 1.2^0$, i.e. 0111. Thus the truth table for the connective called 7 is

$$
\begin{array}{c}
7 \\
\hline
1 \\
1 \\
1 \\
0 \\
\end{array}
$$

Under this naming, the connectives $\wedge$, $\vee$ and $\Rightarrow$ with truth tables

$\wedge$	, $\vee$	and $\Rightarrow$
0	0	1
0	1	1
0	1	0
1	1	1

respectively are connectives numbered 8, 14 and 11 respectively. Which numbered connective is ⇔?

The reason why we single out $\sim$, $\wedge$, $\vee$ and $\Rightarrow$ is simply that these appear to be the natural connectives as far as homo sapiens is concerned. It is true that all $4 + 16 = 20$ connectives can be defined in terms of $\sim$, $\wedge$, $\vee$ and $\Rightarrow$ but this is also true of just $\sim$, $\wedge$ and $\vee$ since $(p \Rightarrow q)$ has the same truth table as $(\sim p \vee q)$ thus showing that $\Rightarrow$ can be defined in terms of $\sim$ and $\wedge$, see Exercises 2, Question 1, part (iii). Indeed, connectives 1 and 7 each have the property that all 19 other connectives can be defined in terms of the one connective. Number 1 is true only when both the statements it connects are false. We read "neither p nor q" and write "$p \downarrow q$". It is called the *joint denial* or *nor* connective. Number 7 is known as *Sheffer's connective*. Read "not both p and q". It is written "$p|q$" also called the *nand* connective; "$p|q$" is false only when p and q are both false. These two connectives find a use in electrical circuit design, since if one builds a circuit which behaves analogously to "$p|q$" then all the circuits which behave like the other connectives can be built up from several of these "nand" circuits.

Exercises 4

1. Given that all 20 connectives can be defined in terms of $\sim$, $\wedge$ and $\vee$, prove that all 20 connectives can be defined in terms of $\sim$ and $\vee$ by showing that $\wedge$ can be defined in terms of $\sim$ and $\vee$.
2. Show that $\sim$ and $\vee$ can be defined in terms of the Sheffer connective $|$.
3. Show that $\sim$ and $\vee$ can be defined in terms of the "nor" connective $\downarrow$.
4. Using connective number n, let us write $p(n)q$ to denote that p is connected to q by n. Explain why $\sim(p(n)q) \Leftrightarrow p(15 - n)q$.

Answers 4

The connective ⇔ is number 9.

1. $(p \wedge q) \Leftrightarrow \sim(\sim p \vee \sim q)$ is a tautology; i.e. $(p \wedge q)$ and $\sim(\sim p \vee \sim q)$ have the same truth table.
2. $\sim p \Leftrightarrow p \downarrow p$ and $p \vee q \Leftrightarrow (p \downarrow q) \downarrow (p \downarrow q)$.
3. $\sim p \Leftrightarrow p|p$ and $p \vee q \Leftrightarrow (p|p)|((p|p)|q)$.

4. Let the binary for n be $a_4a_3a_2a_1$, where a_4, a_3, a_2, a_1 are either zero or one. Then the truth table for $p(n)q$ is

$$\begin{vmatrix} a_1 \\ a_2 \\ a_3 \\ a_4 \end{vmatrix}$$

The truth table for $\sim(p(n)q)$ is thus

$$\begin{vmatrix} 1 - a_1 \\ 1 - a_2 \\ 1 - a_3 \\ 1 - a_4 \end{vmatrix}$$

since, for example, if $a_1 = 0$ then $1 - a_1 = 1$ while if $a_1 = 1$ then $1 - a_1 = 0$.

On the other hand, the binary for $15 - n$ is

$$1111 - a_4a_3a_2a_1 = (1 - a_4)(1 - a_3)(1 - a_2)(1 - a_1)$$

as a four figure binary number. Thus the truth table for $p(15 - n)q$ is

$$\begin{vmatrix} 1 - a_1 \\ 1 - a_2 \\ 1 - a_3 \\ 1 - a_4 \end{vmatrix}$$

identical to that of $\sim(p(n)q)$.

CHAPTER 3

A DESCRIPTION OF PROOF

§1 Rules of Inference

Take an inexhaustible list $p, q, r, \ldots$ of symbols, which we call *statement variables* or elementary statements. Take a list $\sim$, $\vee$, of symbols which we call *logical constants* together with brackets for punctuation. We can regard these symbols as the vocabulary of a language. We define $\wedge$, saying that $P \wedge Q$ is an abbreviation for $\sim(\sim P \vee \sim Q)$. We observe that $P \wedge Q \Leftrightarrow \sim(\sim P \vee \sim Q)$ is a tautology. Similarly $P \Rightarrow Q$ is an abbreviation for $\sim P \vee Q$. Sentences of the language are formed according to definition 2 of the preceding chapter. We have isolated certain sentences of this language, the tautologies, which are the statements of the language which are always true. The next step is to graft onto the language a deductive system. This involves specifying an algorithm which contains a decision process. The function of the algorithm is to decide whether a given statement is a tautology. When we exhibit the application of the algorithm to a statement, and the algorithm decides that the statement is a tautology, we say we have proved the tautology and the application of the algorithm is offered as the proof of the tautology.

We have already given a suitable algorithm using truth tables, for deciding whether a given statement is a tautology. Although this method suffices for the language of statements, it will not generalize for more useful languages. We therefore indicate two other deductive systems, either of which will decide when a statement is a tautology. These agree more closely with what we normally think of as proof.

A proof of a sentence R consists of a list of statements with the following properties.

(i) The first four statements in the list are $(p \vee p) \Rightarrow p$, $p \Rightarrow (p \vee q)$, $(p \vee q) \Rightarrow (q \vee p)$, $(p \Rightarrow q) \Rightarrow (r \vee p \Rightarrow r \vee q)$.

(ii) All the statements in the list, except the first four, have been added to the list by virtue of one of the following two rules.

(a) If P and Q are statements such that P and $P \Rightarrow Q$ are in the list, then we may add Q to the list.

(b) If P is a statement in the list, which contains an elementary statement s in it, then we may take any statement Q and obtain a statement P' by substituting Q in place of each occurence of s in P and we may add P' to the list.

(iii) The statement R occurs in the list.

The first four statements are described as axioms. The rules (a) and (b) are described as rules of inference, although (b) is sometimes described as a rule of substitution. The rule (a) is called *modus ponens*. We observe that the rules generate new statements from statements which are already on the list; the axioms give us some statements to start with. Also, we observe that truth or falsity are not mentioned. A statement is called provable if there exists a proof of it. From the definition of proof it follows that the axioms are provable. We may envisage the collection of provable statements. It is not obvious that this consists of just the tautologies, although this is the case. We can however see that a provable statement must be a tautology. Firstly, we may verify by truth tables that the axioms are tautologies. Secondly the rules (a) and (b) are such that if the statements they are given are tautologies then so is the statement generated.

EXAMPLE 4   A proof of $p \vee \sim p$.

1. $(p \vee p) \Rightarrow p$
2. $p \Rightarrow (p \vee q)$
3. $(p \vee q) \Rightarrow (q \vee p)$
4. $(p \Rightarrow q) \Rightarrow (r \vee p \Rightarrow r \vee q)$
5. $p \Rightarrow p \vee p$ by (b), p replaces q in 2.
6. $((p \vee p) \Rightarrow q) \Rightarrow ((r \vee (p \vee p)) \Rightarrow r \vee q)$ by (b), $p \vee p$ replaces p in 4.
7. $(p \vee p \Rightarrow p) \Rightarrow ((r \vee (p \vee p)) \Rightarrow r \vee p)$, by (b), p replaces q in 6.
8. $(p \vee p \Rightarrow p) \Rightarrow ((\sim p \vee (p \vee p)) \Rightarrow \sim p \vee p)$, by (b), $\sim p$ replaces r in 7.
9. $(p \vee p \Rightarrow p) \Rightarrow ((p \Rightarrow p \vee p) \Rightarrow (p \Rightarrow p))$, remembering that $P \Rightarrow Q$ was short for $\sim P \vee Q$.
10. $(p \Rightarrow p \vee p) \Rightarrow (p \Rightarrow p)$, by (a) using 1 and 9.
11. $p \Rightarrow p$, by (a) using 5 and 10.
12. $\sim p \vee p$, expanding the abbreviation.
13. $(\sim p \vee q) \Rightarrow (q \vee \sim p)$ by (b), $\sim p$ replaces p in 3.
14. $(\sim p \vee p) \Rightarrow (p \vee \sim p)$ by (b), p replaces q in 13.
15. $p \vee \sim p$ by (a) using 12 and 14.

Proofs in this system of deduction tend to be very long. We can shorten these proofs in practice by allowing ourselves to add to a proof list any statement which has already been proved. This avoids the necessity of repeating the proof of a statement every time we need that statement in the proof of another statement. The real problem however is that there is no machinery in this deductive system for the introduction of unproved statements as hypotheses into a proof. We give next a broad description of hypotheses and rules of inference without reference to any particular language.

The word *theorem* generally refers to a statement which is always true. All mathematical theorems take the form of an implication; that is they state that if certain statements, called the hypotheses of the theorem, are true then certain other statements, called the conclusions of the theorem, are also true. Let us consider a theorem whose hypotheses are statements P_1 and P_2 and whose conclusion is R_1. The statement of this theorem might read "If P_1 and P_2 then R_1", which we may express as "$(P_1 \wedge P_2) \Rightarrow R_1$"; a proof of the theorem is a proof that the statement of the theorem is always true.

The proof of such a theorem will consist of a list of statements $P_1, P_2, Q_1, Q_2, Q_3, Q_4, R_1$. When we construct the proof, we begin by writing the hypotheses P_1 and P_2. The statement Q_1 is obtained from P_1 and P_2 by means of a rule of inference. The statement Q_2 is obtained from P_1, P_2, and Q_2 by another rule of inference. At each stage of the proof, a statement may be added to the list using any of the statements already on the list together with a rule of inference.

A rule of inference may be imagined as a machine with an input and an output. It accepts as input one or more statements, which are constrained to be of a certain form. It emits as output usually just one statement. When a rule of inference is used in a proof, the input statements are obtained from the proof list as it stands at the time of using the rule. The output statement is then added to the list.

There are in fact three ways in which a statement can get into a proof. Either it is a hypothesis, or it is a statement already proved to be always true, or it is the output of a rule of inference whose input consists of statements already in the proof list.

The essential feature of a rule of inference is that if the input statements are all true then so is the output statement. It must be firmly impressed however that if one or more of the input statements is false, it does *not* follow that the output statement is false; it can yet be true, but it need not be true.

Suppose we have a machine which accepts a statement P as input and outputs a statement Q. According to the preceding paragraph this machine constitutes a correct rule of inference provided that it is always true that "If P is true, then Q is true", that is the statement $P \Rightarrow Q$ is always true. Thus in order to prove that our machine is a correct rule of inference we need to prove that $P \Rightarrow Q$ is always true, i.e. that $P \Rightarrow Q$ is a *theorem*.

In general, in order to prove that a certain rule of inference is correct, we need to prove that the associated implication is a theorem. In order to prove this, we may have need of certain other rules of inference which we have already proved. Conversely, if we have proved a theorem, such as our original theorem $(P_1 \wedge P_2) \Rightarrow R_1$, then we have proved a rule of inference; here we have proved that a machine which takes as input $P_1 \wedge P_2$ and outputs R_1 is a correct rule of inference. In order to use a theorem as a rule of inference, the statement of the theorem must be expressed as an implication. With this proviso, there is no difference between proving a theorem and proving that a rule of inference is correct, or between using a rule of inference and "applying" a theorem.

When writing down a proof list, it is usual to omit mentioning most of the rules of inference used to extend the list, hoping that the reader will guess which rule of inference is being utilized. This not infrequently means that the reader's comprehension of the proof is dependent upon his familiarity with the sort of rules of inference likely to be used. When we say that a proof is wrong, we do not necessarily mean that the conclusion of the proof is false. We mean that a rule has been used which is not a correct rule of inference.

A deductive system for the language of statements which uses only rules of inference and has no axioms uses the following methods of extending proof lists.

	Input	*Output*
1.	A, B	$A \wedge B$
2.	A	$A \vee B$
3.	B	$A \vee B$
4.	$A \wedge B$	A
5.	$A \wedge B$	B
6.	$A, A \Rightarrow B$	B
7.	$\sim(\sim A)$	A

In addition, there are rules of inference which utilize sub-proofs based on hypotheses. We write $P_1, P_2 \vdash R_1$ to indicate the existence of a proof list which begins with P_1 and P_2 as hypotheses and is extended by any of the rules of inference until it includes R_1.

8. If $A \vdash B$, that is we have a proof of B under the hypothesis A, and $\sim B$ occurs in the proof list we are extending, then we may add $\sim A$ to the list.

9. If $A \vdash C$ and $B \vdash C$ and $A \lor B$ occurs in the proof list we are extending then we may add C.

10. If $A \vdash B$ then we may add $A \Rightarrow B$.

We observe that in order to start a proof list we must supply hypotheses to get some statements for our rules of inference to work on, for we have no axioms. The only exception is the last rule of inference. This requires a proof of B, which will include the hypothesis A, to exist in order to justify the addition of $A \Rightarrow B$ to the proof list currently being extended—but the proof list being extended is not required to include any statement. A statement is proved in this system if there is a proof list which includes the statement but has no hypotheses in it; only 10 allows this possibility. Thus the provable statements are essentially implications. It is convenient in this system to regard $\sim$ and $\Rightarrow$ as the basic connective and $\land$, and $\lor$ as defined in terms of $\sim$ and $\Rightarrow$.

EXAMPLE 5   To prove $(p \lor p) \Rightarrow p$.

The proof $p \vdash p$ consists of the single hypothesis p and does not need to be extended, it already includes p. The proof $p \lor p \vdash p$ is obtained as follows.

Hypothesis, $p \lor p$. Then by 9, $p \lor p$ occurs in this list and we have the proof $p \vdash p$ whence we can add p to this list. Since we have the proof $p \lor p \vdash p$, by 10, we can start a proof list, without hypotheses, with the statement $p \lor p \Rightarrow p$. This concludes the proof of $p \lor p \Rightarrow p$.

This specific system is offered for comparison with the more general description of the proof of implications which follows.

§2 Direct Proof

Let us return to our theorem $(P_1 \land P_2) \Rightarrow R_1$ and its proof P_1, P_2, Q_1, Q_2, Q_3, Q_4, R_1, and ask why it is that the proof demonstrates that the theorem is always true. We do not know whether P_1 and P_2 are true or false, but if one of them is false then $P_1 \land P_2$ is false so that the implication $(P_1 \land P_2) \Rightarrow R_1$ is true according to the truth table of $\Rightarrow$. If P_1 and P_2 are both true, then we must appeal to the proof list. Notice that although the proof list only applies when P_1 and P_2 are true, a proof list must always be provided in order to prove $(P_1 \land P_2) \Rightarrow R_1$ because we do not know the truth values of P_1 and P_2 and so it is essential that this case be covered.

Now, if P_1 and P_2 are both true then every statement in the proof list must be true, for consider how a statement can get into the proof list. Either it is a hypothesis, i.e. it is either P_1 or P_2 and is therefore assumed true in the case we are considering; or it is a statement already shown to be always true; or it is the output of a rule of inference. But a rule of inference whose input statements are true must generate true statements. Hence if Q_1 is obtained using P_1 or P_2 as the input of a rule of inference then since P_1 and P_2 are true, the output statement Q_1 must be true. Similarly if Q_2 is obtained using P_1, P_2 or Q_1 as the input of a rule of inference then since the input statements, i.e. P_1, P_2 or Q_1, are all true then the output statement Q_2 must be true. Thus at each stage in the construction of the proof list, the statements already on the list are all true so that if we use them as the input of a rule of inference, the output of the rule which we add to proof list must be true and so we can repeat the process. The conclusion R_1 at the end of the proof list is true therefore, as indeed is every statement in the list. Thus R_1 is true and the truth table of $\Rightarrow$ assures us that the theorem $(P_1 \wedge P_2) \Rightarrow R_1$ is true, for we have shown that if the truth value of $(P_1 \wedge P_2)$ is 1, then the truth value of R_1 is 1.

This method of proving the theorem $(P_1 \wedge P_2) \Rightarrow R_1$ is called a *direct proof*. Compare the remarks on direct proof and implication in Chapter 2, §2.

§3 Indirect Proof

Let us suppose that we are investigating a statement of the form $P \Rightarrow Q$ to see whether it is true or false. We may show, using truth tables, that the following statement is a tautology.

$$(P \Rightarrow Q) \Leftrightarrow (\sim Q \Rightarrow \sim P)$$

This means that if $P \Rightarrow Q$ is true, then so is $\sim Q \Rightarrow \sim P$ while if $P \Rightarrow Q$ is false, then so is $\sim Q \Rightarrow \sim P$. Thus our investigation of $P \Rightarrow Q$ may take the form of an investigation of $\sim Q \Rightarrow \sim P$, since in establishing the truth or falsity of $\sim Q \Rightarrow \sim P$ we shall also establish the truth or falsity of $P \Rightarrow Q$.

Let us suppose next that our investigations lead us to believe that $\sim Q \Rightarrow \sim P$ is true. We must now supply a proof of this statement. The hypothesis of the statement is $\sim Q$ and the conclusion of it is $\sim P$. Thus a direct proof will take the form:

$$\sim Q, Q_1, Q_2, Q_3, \sim P$$

say, where we have presumed there to be three intermediate steps Q_1, Q_2, Q_3.

This proof is a direct proof of $\sim Q \Rightarrow \sim P$. But $(\sim Q \Rightarrow \sim P) \Leftrightarrow (P \Rightarrow Q)$, so that this proof may be thought of as proving $P \Rightarrow Q$. This method of proving $P \Rightarrow Q$ is called the *indirect method*. Compare the remarks on indirect proof and implication in Chapter 2, §2.

Notice that the proof begins by taking $\sim Q$ to be true, i.e. Q to be false, and proceeds to demonstrate that $\sim P$ is true, i.e that P is false. Overtly therefore, the indirect method of proving $P \Rightarrow Q$ differs greatly from the direct method. In fact, an indirect proof of $P \Rightarrow Q$ is a direct proof of the equivalent statement $\sim Q \Rightarrow \sim P$.

EXAMPLE 6   If a is a natural number and b is a natural number then if $a + b = 2$ then $a = 1$ and $b = 1$. By a natural number we mean a counting number, i.e. one of 1, 2, 3, ...

We have to prove the implication $(a + b = 2) \Rightarrow (a = 1) \wedge (b = 1)$. We use the indirect method, which is in fact the direct method of proving the equivalent statement $\sim((a = 1) \wedge (b = 1)) \Rightarrow (a + b = 2)$. Thus we begin

1. $\sim((a = 1) \wedge (b = 1))$

   "It is false that both $a = 1$ and $b = 1$".

2. $\sim(a = 1) \vee \sim(b = 1)$

   "$a \neq 1$ or $b \neq 1$"

We have obtained 2 from 1, using the fact that $\sim(P \wedge Q) \Leftrightarrow \sim P \vee \sim Q$ is a tautology, where P in our case is $a = 1$ and Q is $b = 1$. Thus if $\sim(P \wedge Q)$ is true, so is $\sim P \vee \sim Q$ and the rule of inference which accepts $\sim(P \Rightarrow Q)$ as input and outputs $\sim P \vee \sim Q$ is a correct rule of inference.

3. $(a > 1) \vee (b > 1)$

Here we have used a rule of inference, a machine which accepts as input any statement of the form "x is a natural number and $x \neq 1$" and generates the statement "$x > 1$". This rule does not depend on verifying a tautology even though

$$(x \text{ is a natural number and } x \neq 1) \Rightarrow (x > 1)$$

is always true. For the moment, we hope that the reader's intuition is satisfied that this is a correct rule of inference. This rule of inference tells us that if $a \neq 1$ is true, so is $a > 1$ and if $b \neq 1$ is true, so is

$b > 1$. Since 2 assures us that at lease one of $a \neq 1$ and $b \neq 1$ is true, at least one of $a > 1$ and $b > 1$ is true, whence 3.

4.
$$a + b \geqslant 3$$

If one of two numbers is greater than 1, i.e. at least 2, and the other is at least 1 then their sum is at least 3.

5.
$$a + b \neq 2$$
$$\text{i.e. } \sim(a + b = 2).$$

If a number is greater or equal to 3 then it is not equal to 2. This concludes the proof.

Several of the rules of inference used here are not tautologies, but rely for their truth upon properties of the ordering and addition of natural numbers. We shall not examine these statements formally yet.

§4 Proof by Contradiction

Let us suppose that we have a proof list P, Q_1. Q_2, Q_3, R and that further the statement R is known to be false. The statement P is the hypothesis while Q_1, Q_2, Q_3 and R have been added to the list either because they have already been proved to be true or else under the auspices of a rule of inference.

Now, were P true then would every statement in the proof list be true. But R is false and therefore P must be false. Thus we have a method of proving that a statement is false, which we can use to prove that its negation is true.

Suppose we wish to prove that a statement T is true. We begin a proof list with the statement $\sim T$ and extend it by the usual means until a statement occurs in the list which is known to be false. The list might be $\sim T$, Q_1, Q_2, Q_3, R where R is known to be false. Again, if $\sim T$ were true then would all the statements in the list be true, which is not the case. Hence $\sim T$ must be false and so T must be true. For this method to work, it is not necessary to know which statement is false, although we can in fact always produce a statement which is known to be false.

Consider the proof list $\sim T$, Q_1, Q_2, Q_3, R, Q_4, $\sim R$. Clearly one of the two statements R or $\sim R$ must be false; were $\sim T$ true then all the statements would be true; hence $\sim T$ is false. We could extend the list one step by adding $R \wedge \sim R$ using the rule of inference which says that if X and Y are true then $X \wedge Y$ is true. Truth tables assure us that $R \wedge \sim R$ is always false.

It is quite possible to obtain a proof by contradiction of T whose list is of the form $\sim T$, Q_1, Q_2, Q_3, T. This shows that if $\sim T$ is true then so is T, evidently not possible. Hence $\sim T$ is false.

Suppose we wish to prove a statement of the form $P \Rightarrow Q$ by contradiction. The first steps of the proof list would probably be a condensed version of $\sim(P \Rightarrow Q)$, $\sim(P \Rightarrow Q) \Rightarrow (P \wedge \sim Q)$, $P \wedge \sim Q$, $(P \wedge \sim Q) \Rightarrow P$; $(P \wedge \sim Q) \Rightarrow \sim Q$, P, $\sim Q$.

Here, we added $\sim(P \Rightarrow Q) \Rightarrow (P \wedge \sim Q)$ and $(P \wedge \sim Q) \Rightarrow P$ and $(P \wedge \sim Q) \Rightarrow \sim Q$ to the list because they are tautologies. We used the rule of inference $(X \wedge (X \Rightarrow Y)) \Rightarrow Y$, known as modus ponens, to obtain $P \wedge \sim Q$, by taking $\sim(P \Rightarrow Q)$ for X and $P \wedge \sim Q$ for Y. The same rule was used to obtain P from $P \wedge \sim Q$ and $(P \wedge \sim Q) \Rightarrow P$ and also to obtain $\sim Q$ from $P \wedge \sim Q$ and $(P \wedge \sim Q) \Rightarrow \sim Q$.

Thus beginning the proof with $\sim(P \Rightarrow Q)$ involves taking P and $\sim Q$ as hypotheses. In a sense this means assuming that P is true and Q is false. Thus a proof of $P \Rightarrow Q$ by contradiction involves giving P the truth value 1 and Q the truth value 0 and then deducing a contradiction. Compare the remarks on contradiction and implication in Chapter 2 §2.

EXAMPLE 7   Prove that "The only prime natural number which is even is 2". By an even natural number we mean a natural number which is exactly divisible by 2 and by a prime natural number we mean a natural number other than 1 which is exactly divisible only by 1 and itself.

We begin the proof list with the negation of the statement we wish to prove.

1.   2 is not the only prime natural number which is even.
2.   There is a natural number r of which the following three statements are true.
2a. $r \neq 2$.
2b. r is prime.
2c. r is even.
3. r is exactly divisible by 2 (from 2c by definition of "even").
4. r is exactly divisible only by 1 and r (from 2b by definition of "prime").
5. $2 = 1$ or $2 = r$ (from 3 and 4).
6. $2 = 1$, using rule of inference

$$X \wedge (\sim X \vee Y) \Rightarrow Y$$

in which X is the statement 2a and Y the statement "$2 = r$". Notice that $(\sim X \vee Y)$ is then statement 5. That this rule of inference is correct may be verified by proving by truth tables that $X \wedge (\sim X \Rightarrow Y) \Rightarrow Y$ is a tautology.

Now, 6 is known to be false, so 1 must be false and therefore the negation of 1, which is the original statement to be proved is true.

§5 Use of Equivalent Statements

Let us suppose that we wish to prove a statement P and that although we do not know whether P itself is true or false, we do know that $P \Leftrightarrow Q$ is true. If we can find a proof of Q then we can extend this proof to a proof of P using $P \Leftrightarrow Q$, since Q is true if and only if Q is true. This process is advantageous if Q is easier to prove than P. In §3 it was explained that instead of proving $P \Rightarrow Q$, it is sometimes more convenient to prove the equivalent statement $\sim Q \Rightarrow \sim P$.

For a more striking example let us consider proving a statement of the form $P \Rightarrow (Q \vee R)$. Now $Q \vee R$ is true when either Q is true or R is true, but although we need to show that at least one of these statements is true we do not need to show which. A direct proof of $P \Rightarrow (Q \vee R)$ as it stands is liable to be trying not only to prove, under the hypothesis P, that at least one of Q and R is true but also which, and this may not be possible. Also, proofs have to be engineered towards a known conclusion and this is more easily done if the conclusion has a simple form. For both the above outlined reasons it is often preferable not to give a direct proof of $P \Rightarrow (Q \vee R)$, but rather a direct proof of one of the equivalent statements $(P \wedge \sim Q) \Rightarrow R$ or $(P \wedge \sim R) \Rightarrow Q$. Such a proof has hypothesis say $P \wedge \sim Q$ and conclusion R and intuitively one has more hypothesis and less conclusion than in proving $P \Rightarrow Q \vee R$ so one might expect it to be easier to find a proof.

We might try to find an indirect proof of $P \Rightarrow (Q \vee R)$, that is a direct proof of $\sim(Q \vee R) \Rightarrow \sim P$. This is not in a convenient form yet. In proving $(P \wedge \sim Q) \Rightarrow R$ we can separate the hypothesis $P \wedge \sim Q$ into two hypotheses P and $\sim Q$ and such a proof would in fact often begin "Let P be true and let Q be false". But we cannot separate the hypothesis $\sim(Q \vee R)$ into two hypotheses, the one about Q and the other about R. However, $\sim(Q \vee R) \Leftrightarrow \sim Q \wedge \sim R$ is a tautology. Therefore, instead of taking $\sim(Q \vee R)$ as our hypothesis we may take $\sim Q \wedge \sim R$ which we can separate into two hypotheses, $\sim Q$ and $\sim R$. The first line of the proof list of $(\sim Q \wedge \sim R) \Rightarrow \sim P$ would read "Let Q and R both be false" or "Let $\sim Q$ and $\sim R$ be true". Thus we have replaced $P \Rightarrow (Q \vee R)$ by the equivalent statement $(\sim Q \wedge \sim R) \Rightarrow \sim P$.

We have, incidentally, given an example of an important property of equivalent statements. We took a part of the statement $\sim(Q \wedge R)$ $\Rightarrow \sim P$, namely the $\sim(Q \vee R)$ part, and replaced it with $(\sim Q \wedge \sim R)$ claiming as justification that $\sim(Q \vee R)$ is equivalent to $\sim Q \wedge \sim R$, i.e. that $\sim(Q \vee R) \Leftrightarrow (\sim Q \wedge \sim R)$ is a tautology. In general, if a compound statement T contains a simple statement P as part of it and if $P \Leftrightarrow Q$ then we can replace P by Q in the statement T, obtaining thereby a statement T' and T' will be equivalent to T, i.e. $T \Leftrightarrow T'$ will be true. Thus, for example, let T be

$$(P \Rightarrow Q) \wedge (Q \Rightarrow R).$$

We know that $(P \Rightarrow Q) \Leftrightarrow (\sim Q \Rightarrow \sim P)$. Thus we make the substitution in T to obtain as T' the statement $(\sim Q \Rightarrow \sim P) \wedge (Q \Rightarrow R)$. To see that $T \Leftrightarrow T'$ we must show that T and T' have the same truth table. Consider running a truth table on T and T'. In running the table on T, we shall have a column for $P \Rightarrow Q$ and in the table on T' we shall find instead a column for $\sim Q \Rightarrow \sim P$. However, these two columns will be identical since $(P \Rightarrow Q) \Leftrightarrow (\sim Q \Rightarrow \sim P)$. Hence the truth tables of T and T' will be the same.

We may formulate a rule of inference which accepts as input any statement T, say, which has a simple statement X as part of it, together with a statement $X \Leftrightarrow Y$ which is true, and outputs a statement T' obtained by replacing X in T by Y.

If the following exercises, the reader is invited to supply proofs and to analyse these proofs in terms of the descriptions in the preceding chapter as far as possible.

Exercises 5

1. Prove that "If the square of a natural number is odd then the number itself is odd". An odd natural number is one which is not exactly divisible by 2.

2. Prove that "If p is a prime natural number and a and b are natural numbers and p divides ab exactly then either p divides a exactly or p divides b exactly". Give an example of three natural numbers p, a and b such that p is not prime and p divides ab exactly but p does not divide either a or b exactly.

3. An integer is either a natural number or zero or a negative natural number, i.e. the integers are 0, $+1$, -1, $+2$, -2, $+3$, -3, ... Let b be an integer. Prove that (i) $b^2 \neq -1$ and (ii) $b^2 \neq 3$ and (iii) $b^2 \neq \frac{1}{2}$.

4. Let i be a number such that $i^2 = -1$. The following rules of inference apply to the relationship $<$ between numbers.

   If a, b and c are numbers, then

   Ia $(a < b) \Rightarrow (a + c < b + c)$
   Ib $(a < b) \Rightarrow (a - c < b - c)$
   IIa $((c > 0) \wedge (a < b)) \Rightarrow (ac < bc)$
   IIb $((c < 0) \wedge (a < b)) \Rightarrow (ac > bc)$.

   Use these rules of inference to prove that

   (i) $(i > 0) \Rightarrow -1 > 0)$
   (ii) $(i < 0) \Rightarrow -1 > 0)$
   (iii) $(i = 0) \Rightarrow (1 = 0)$.

   What can we conclude from (i), (ii) and (iii) about the statement $(i > 0) \vee (i < 0) \vee (i = 0)$? What can we conclude about the number i and the relationship $<$?

Answers 5

1. Suppose the number is not odd.
   Then it is even.
   Therefore is has a factor 2.
   Therefore its square has a factor 2.
   Therefore its square is even.
   i.e. its square is not odd.

2. It is a well known theorem that a natural number can be uniquely expressed as a product of prime factors. Thus a prime divides a number if and only if it occurs as a prime factor in the expression of the number. Let p be prime and let p divide ab and let p not divide a. The expression for ab may be obtained by multiplying the expression for a by the expression for b. Now p occurs in the expression for ab but not in the expression for a. Hence it must occur in the expression for b. Hence p divides b exactly.
   Let $p = 6$, $a = 2$, $b = 3$. The p divides ab but does not divide either a or b.

3. (i) It is true that either $b > 0$ or $b = 0$ or $b < 0$. If $b > 0$ then $b^2 > 0$; if $b = 0$ then $b^2 = 0$ and if $b < 0$ then $b^2 > 0$. Hence $b^2 \geqslant 0$. Hence $b^2 \neq -1$.
   (ii) Now $1^2 = 1$ and $2^2 = 4$ and if $b > 2$ then $b^2 > 4$. Hence $b^2 \neq 3$.
   (iii) Now $b \geqslant 1$ so $b^2 \geqslant 1 > \frac{1}{2}$. Hence $b^2 \neq \frac{1}{2}$. Alternatively, if b

is a natural number so is b^2, but $\frac{1}{2}$ is not a natural number so $b^2 \neq \frac{1}{2}$.

4. (i) If $i > 0$ then by IIa, $i^2 > 0 . i = 0$ so $-1 > 0$.

(ii) If $i < 0$ then by IIb, $i^2 < 0 . i = 0$ so $-1 > 0$.

(iii) If $i = 0$ then $i^2 = 0 . i = 0$ so $-1 = 0$ so $-1 . -1 = 0 . -1$ whence $1 = 0$.

Now, (i) shows that $(i > 0)$ is false since $-1 > 0$ is known to be false. Similarly (ii) and (iii) show that $(i < 0)$ and $(i = 0)$ are false. Hence $(i > 0) \vee (i < 0) \vee (i = 0)$ is false.

By a real number we mean, loosely speaking, a decimal. Now we are familiar with the fact that any statement obtained from the expression $(x > 0) \vee (x < 0) \vee (x = 0)$ by replacing x by a real number will be a true statement. But if we replace x by the number i we obtain a false statement. We might conclude that whatever sort of object i may be, it is not a real number.

CHAPTER 4

PREDICATES

§1 Variables and Parameters

In the preceding chapter we dealt with the language of statements. We have said a great deal about this language, but we can say very little of mathematical interest in the language. In the mathematical examples, we have often lapsed into English in order to express statements and rules of inference too subtle for the statement language to cope with adequately.

Let p, q and r be the statements

"All linear equations can be solved",
"$3x + 1 = 0$ is a linear equation" and
"$3x + 1 = 0$ can be solved" respectively.

Then the compound statement $(p \wedge q) \Rightarrow r$ is evidently true, yet it is not a tautology nor can we analyse the statement further in the statement language. We require a more subtle language in order to analyse p, q and r further and prove theorems like the above. It is at this point that the truth tables fail us as a method of determining whether a statement is true or not.

Consider the four statements

"1 is an even number"
"2 is an even number"
"3 is an even number"
"4 is an even number".

These are all perfectly good statements some of which are true and some false. Evidently there are a good many more similar statements, and we shall never be able to write down all of them. What we can do however is to give a recipe for writing down such statements. We can say that "... is an even number" becomes a statement whenever we fill in the blank space with a natural number.

This idea of constructing recipes or rules is very important since it is the only way we have of dealing with a large or infinite number of objects or statements. In our example, the recipe "gives" us an infinite collection of statements but it does not "give" us this collection in the same sense as we gave the finite collection of four statements in the last paragraph. One must be careful when "giving" an infinite collection of objects by means of a recipe that one knows in what sense the collection is "given". The recipe of the example will not enable us to construct all the statements for which it is a recipe. It will however enable us to construct any finite number of those statements that we wish.

The first improvement to the method of stating the recipe is to fill in the space with a space marker. We shall use "n". Thus our recipe becomes:

(*) "n is an even number" is a statement whenever we write a natural number in the space marked with the letter n.

There are more fundamental reasons for introducing n, but for the moment we simply observe that spaces are difficult to reproduce on paper consistently and also difficult to pronounce.

The letter n is used here as a *variable*. It is a space marker and is *not* itself a natural number. We could decide to use m as a space marker instead of n. The recipe: ["m is an even number" is a statement whenever we write a natural number in the space marked with the letter m] is exactly the same recipe as before.

It is a mistake to take the statement "n is an even number" outside the context of the recipe since then we do not know what objects may be used to fill the space marked n. In mathematics, one always has a collection of objects attached to each variable, often called the *substitution set* of the variable. Any proto-statement which includes a variable should always be somehow attached to a remark which specifies the substitution set of that variable, because a variable is used when we wish to discuss a whole collection of statements and it is important to know exactly what statements are in the collection.

Notice that in the recipe (*) the variable n has no meaning of its own outside the text of the recipe. We are not using n to represent a number which we do not know. Contrast this use of letters with the use of the letter b in the following sentence.

"Let b be the positive number whose square is 2".

Here we introduce b as a name for a number; we perhaps do not know what this number is as a decimal. Nevertheless, there is no substitution set for b. It is used as the name of a number, not as a space marker. The sentence "$b^2 = 2$" is a statement, indeed a true statement. The sentence "$b > 2$" is a statement, indeed a false statement. A sentence like "$b^5 > 6$" is a statement since it is certainly either true or false even though we may not know which.

We describe the letter b as a *parameter*. Letters should always be properly introduced. In the case of a variable, this office is performed by specifying the substitution set of the variable. Parameters are introduced by stating what sort of object the parameter is. There are two problems associated with introducing parameters, both of which were avoided by our example. In an introduction "Let b be such and such an object" it is possible that there is no such object. For example "Let b be a natural number whose square is 2". Such a definition of b yields the statement "b is a natural number whose square is 2" which is false. Thus introducing a parameter b involves an assumption, namely the assumption that there is indeed a suitable object to be called b.

The second problem is that there may be more than one candidate for the honour of being called b. Thus "Let b be a natural number". In this event we take the view that b is the name of a particular object but the introduction is not complete in that it is not sufficiently good to enable us to single out b from other objects which the description also fits. In spite of the fact that we have not described b completely, we can make up sentences involving b. These sentences will be statements in that they will be either true or false; of course, we may not be able to say which because b has not been completely described.

By way of an example of these two different uses of letters, contrast the two equations.

(1) $(x + 1)^2 = x^2 + 2x + 1$
(2) $x^2 + 3x + 2 = 0.$

The use of x is not clear since this letter has not been introduced properly. We shall now use (1) and (2) in sentences, complete with an introduction to x.

"Consider the statements which result from filling the space marked x in the proto-statement $(x + 1)^2 = x^2 + 2x + 1$ by real numbers". Here is an example of x used as a variable. We are considering a whole collection of statements. The interest in considering this collection lies in the fact that every statement in the collection is true.

Again "Let x be a real number and let $(x + 1)^2 = x^2 + 2x + 1$". Here is x used as a parameter, but the second part of the description of x, that it should satisfy the equation $(x + 1)^2 = x^2 + 2x + 1$, is of no interest in view of the fact that every real number satisfies this description.

Again "Consider the collection of statements which result from filling the space marked x in the proto-statement $x^2 + 3x + 2 = 0$ by real numbers". Here is x used as a variable to describe a collection of statements. However, this collection is not of great significance. Some of the statements in the collection are true, some are false. If we say that the true statements are in fact $(-1)^2 + 3(-1) + 2 = 0$ and $(-2)^2 + 3(-2) + 2 = 0$ we still seem to be missing the point. Let us alter our usage of x from the variable use to parametric use.

"Let x be a real number and let $x^2 + 3x + 2 = 0$". The description of x, that it should satisfy the equation $x^2 + 3x + 2 = 0$, is very discerning since the only real numbers which satisfy the description are -1 and -2. This description of x may therefore be put in a more direct form, namely "Let x be either -1 or -2". This does not mean that x is in some sense both -1 and -2 at the same time. It means that x is one or the other, but the introduction to x is incomplete so we are unable to say which.

We observe briefly that in the case of (1) it was more fruitful to regard x as a variable, while in the case of (2) it was more fruitful to regard x as a parameter, although x can in fact play either role in each case.

We may mention a further distinction between variables and parameters which occurs in proofs. A sentence such as (*) gives the proto-statement and introduces the variable by giving its substitution set. This use of n is contained within the sentence even though the sentence may be part of a larger proof; the introduction in (*) applies only to the use of n in the proto-statement given in (*). The letter n is available for use in other sentences of the proof as a space marker and may be used in a different proto-statement and given a different introduction in such other sentences. This can cause confusion, because mathematicians are in the habit of introducing a variable and allowing the introduction to serve for several proto-statements; more confusing is the fact that when they do this they introduce the variable as if it were a parameter. Thus one commonly finds such examples as the following.

Let n be a natural number. Then

$$1 + 2 + 3 + 4 + \ldots + n = \tfrac{1}{2}n(n + 1).$$

Thus putting

$$n = 5, 1 + 2 + 3 + 4 + 5 = \tfrac{1}{2} \cdot 5 \cdot 6$$

and putting

$$n = 4, 1 + 2 + 3 + 4 = \tfrac{1}{2} \cdot 4 \cdot 5.$$

This is not entirely consistent, since if n is a natural number it cannot possibly be both 4 and 5 neither is any justification given for the belief that it is either. What is meant is the following.

The statements obtained from the proto-statement

$$\text{``}1 + 2 + 3 + 4 + \ldots + n = \tfrac{1}{2}n(n + 1)\text{''}$$

by replacing n by a natural number are all true. Thus replacing n by 5,

$$1 + 2 + 3 + 4 + 5 = \tfrac{1}{2} \cdot 5 \cdot 6 \text{ is true and,}$$

replacing n by 4,

$$1 + 2 + 3 + 4 = \tfrac{1}{2} \cdot 4 \cdot 5 \text{ is true.}$$

It may be as well to state that we are not advocating the second version in preference to the first, but merely that the first be correctly interpreted. Contrast also the introduction of a parameter into a proof. In this case one is not allowed to re-use the parameter. It has been introduced as the name of an object and cannot be re-used as the name of some other object. Neither is it advisable to re-use it as a variable even though this is permissible provided the proto-statement in which it occurs as a space marker does not also contain an occurrence of it in its role as a parameter.

§2 Defining Sets

We have already discussed collections of objects. We shall take a naive view of sets at this point, and define the word *set* to mean simply a collection of objects. The first problem is, how does one specify a set? If we can conceive of a collection of objects then we ought to be able to describe it in such a way as to be able to determine which objects are in the set and which are not. First of all therefore we shall deal with descriptions of objects.

The simplest and most reliable method of describing a thing to someone is to get the three of you in one place and point it out to him. Failing this, you can give him the name of the thing provided that you both have a name for the thing in common. We can describe a set by listing the names of the objects which comprise the set. The names we refer to are grammatically described as proper nouns. It is essential that two different objects should not have the same name. It can and does happen that an object has two or more different names. This is a feature not only of Russian novels but also of mathematics. For example, $\tfrac{1}{1}, \tfrac{2}{2}, \tfrac{3}{3}, \tfrac{4}{4}$ are all different names for the same thing. If we write a list of names describing a set of objects and the list contains two different names for the same object, then the set will contain that object and one of the names will be redundant. We usually list our names one after the other, separated by commas, with the whole list enclosed in curly brackets. Thus the collection comprising the first four natural numbers could be written $\{1, 2, 3, 4\}$ and the set $\{\frac{2}{2}, \frac{4}{4}, 2, 1\frac{2}{4}, 1\frac{2}{3}\}$ is exactly the same collection.

Evidently one cannot describe a set which has infinitely many objects in it by listing the objects. Instead we formulate a description which fits just the objects which we wish to belong to the set.

Consider the statements obtainable from "n is even" by substituting natural numbers for n. We define the collection of even numbers to be the set of natural numbers such that when they replace n in "n is even" the result is a true statement. The usual way of arranging this kind of definition so that it doesn't take up too much room is the following.

Let $\mathbf{N}$ denote the set of all natural numbers, this is the substitution set of the variable n. The set of even natural numbers is then written

$$\{n \in \mathbf{N} \,|\, n \text{ is even}\}.$$

This defines the set of all objects in the substitution set $\mathbf{N}$ which make "n is even" a true statement whenever one of these objects fills the space marked n. We shall reserve curly brackets for defining sets.

The symbol $\in$ is read "is a member of". Whenever A is a set, we write $a \in A$ to express the fact that a is one of the objects in the set A. The objects in A are also called *elements of A* or *members of A*. The notation $n \in \mathbf{N}$ would seem to express the fact that n is a natural number, which it is not, being a variable. This abuse is common, and is actually intended to inform us that $\mathbf{N}$ is the substitution set of the variable n. The vertical stroke merely separates the part of the definition which specifies the variable and its substitution set from the proto-statement and it is conveniently read "such that".

More generally, let us denote by $P(x)$ a proto-statement in which the space, or indeed spaces, are all marked with x. Let S be the substitution set for x. Then the set A of elements of S which make $P(x)$ into a true statement is defined by

$$\{x \in S \,|\, P(x)\}.$$

A proto-statement $P(x)$ is called a *predicate*. A predicate in one variable, or *unary predicate*, is a proto-statement in which only one variable is used. There may well be more than one space in $P(x)$ but all the spaces are marked with x. Thus when we replace x by an element of the substitution set of x, all the spaces in $P(x)$ are filled with that element and the result is a statement. For example, consider the predicate "$1 + 2 + 3 + .. + n = \frac{1}{2}n(n + 1)$" which we may call $P(n)$ for short. Replacing n by 4 we obtain $P(4)$ which is "$1 + 2 + 3 + 4 = \frac{1}{2} \cdot 4 \cdot 5$". There are in fact three spaces in $P(n)$ but each is marked by n.

One question is, if we need a substitution set to define a set then where do the substitution sets come from? We shall beg this question for the moment, remarking only that for substitution sets we shall use well known sets like $\mathbf{N}$, the set of all natural numbers; $\mathbf{Z}$, the set of all integers and $\mathbf{R}$, the set of all real numbers.

Before continuing, we remark that we may always consider a set to be defined using a predicate. It is not necessary to consider separately sets defined by lists and sets defined by predicates. For example the set $\{1, 2, 3, 4\}$ may also be defined as

$$\{x \in \mathbf{N} | (x = 1) \vee (x = 2) \vee (x = 3) \vee x = 4)\},$$

using "$(x = 1) \vee (x = 2) \vee (x = 3) \vee (x = 4)$" as the predicate.

Consider the set A defined by $\{x \in S | P(x)\}$. We introduce a parameter a. Let $a \in A$. Now by our notation, a is an element of A if and only if the result of substituting a for the variable x in $P(x)$ is a true statement. We denote the result of substituting a for x in $P(x)$ by $P(a)$. Thus $P(x)$ is a predicate but $P(a)$ is a statement. Hence $a \in A$ is true if and only if $P(a)$ is true. In symbols

$$(a \in A) \Leftrightarrow P(a).$$

Notice that an infinite set is not "given" by its predicate in quite the same way as a finite set is "given" by a list of its objects. The predicate does not enable us to find all the objects in the set. Indeed, it does not even guarantee to find us even one object in the set. All it does guarantee is that if you present it with an object, then it will tell you whether or not that object is in the set. If you ask whether a is in the set $\{x \in S | P(x)\}$ then $P(x)$ tells you whether it is in the set or not, for if $P(a)$ is true it is in, while if $P(a)$ is false then it is not. It is rather like having an oracle which can only answer yes or no. It cannot tell you where the rainbow ends but only whether or not it ends at the places you suggest. If S is infinite, then you will certainly not be able to hold up each element of S to the scrutiny of $P(x)$.

§3 Compounding Predicates and Sets

Let $P(x)$ and $Q(x)$ be predicates.

Then $\sim P(x)$, $P(x) \wedge Q(x)$, $P(x) \vee Q(x)$ and $P(x) \Rightarrow Q(x)$ are also predicates, for they become statements when x is replaced by an element from the substitution set of x.

We wish to avoid having to associate with each set not only its own predicate but also its own substitution set. This we can do by introducing the concept of a *universal set* which contains all the objects with which any predicate we use might wish to replace its variable. Thus our sets will be defined in the form $\{x \in \mathbf{U} \mid \quad \}$ where $\mathbf{U}$ is the universal set. This may in practice mean adjusting the predicate a little. For example, if we wished to discuss $\{x \in \mathbf{N} \mid x \text{ is even}\}$ and, for the benefit of some other predicate under discussion, the universal set had to be the set $\mathbf{R}$ of reals then we would have to describe our set as

$$\{x \in \mathbf{R} \mid (x \in \mathbf{N}) \wedge (x \text{ is even})\}.$$

The Union of Two Sets. Let A and B be sets of objects. We define the new set, called the *union* of A and B and denoted $A \cup B$, to be the set of all objects which are either in A or in B. Thus to get $A \cup B$, we push the two sets together.

More precisely, if A is defined by

$$\{x \in \mathbf{U} \mid P(x)\}$$

and B is defined by

$$\{x \in \mathbf{U} \mid Q(x)\}$$

then $A \cup B$ is the set defined by $\{x \in \mathbf{U} \mid P(x) \vee Q(x)\}$.

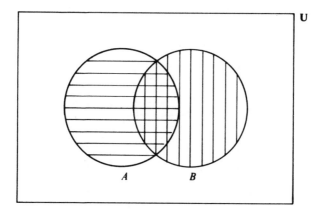

Thus if $a \in A \cup B$ then $P(a) \vee Q(a)$ is true, so either $P(a)$ is true or $Q(a)$ is true. If the former then $a \in A$ and if the latter then $a \in B$.

Let $a \in \mathbf{U}$. Then the following statements are equivalent:

(i) $a \in A \cup B$
(ii) $P(a) \vee Q(a)$
(iii) $(a \in A) \vee (a \in B)$.

i.e. $a \in A \cup B \Leftrightarrow P(a) \vee Q(a) \Leftrightarrow (a \in A) \vee (a \in B)$.

We may draw a *Venn diagram* to illustrate the creation of $A \cup B$ from A and B.

Here, the square and its interior represents the universal set; the horizontally shaded circle the set A and the vertically shaded area the set B. The total shaded area then represents $A \cup B$.

The Complement of a Set. The *complement* of a set A is the set of all elements of the universal set which are not in A. This is denoted A'. It is better to mention the universal set being used and to specify A' as being the complement of A *with respect to the universal set* $\mathbf{U}$.

More precisely, if A is defined by $\{x \in \mathbf{U} | P(x)\}$ then A' is defined to be

$$\{x \in \mathbf{U} | \sim P(x)\}.$$

Let $a \in \mathbf{U}$. Then $a \in A \Leftrightarrow P(a)$ while $a \in A' \Leftrightarrow \sim P(a)$. At this point we introduce the notation $\notin$ for "is not a member of". As a more precise definition, $a \notin A$ is equivalent to $\sim(a \in A)$. Thus $a \in A' \Leftrightarrow a \notin A$.

We may draw a Venn diagram to illustrate A'.

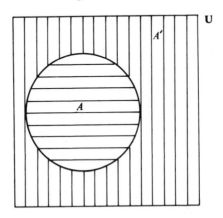

The square and its interior represents the universal set. The set A is represented by the horizontally shaded ring, while A' is the vertically shaded area.

The Intersection of Two Sets. The *intersection* of two sets A and B is the set of all objects which the sets have in common. This set is denoted $A \cap B$.

More precisely, if A is defined by $\{x \in U | P(x)\}$ and B is defined by $\{x \in U | Q(x)\}$ then $A \cap B$ is defined to be $\{x \in U | P(x) \wedge Q(x)\}$.

Thus if $a \in A \cap B$ then $P(a) \wedge Q(a)$ is true; thus both $P(a)$ and $Q(a)$ are true whence $a \in A$ and $a \in B$.

The following three statements are equivalent

(i) $a \in A \cap B$

(ii) $P(a) \wedge Q(a)$

(iii) $(a \in A) \wedge (a \in B)$.

The Venn diagram illustration is as follows.

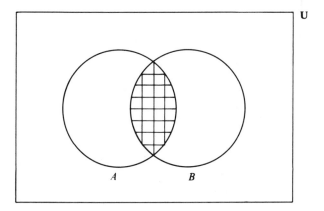

The set $A \cap B$ is represented by the doubly shaded region.

It is of course possible that A and B have no object in common. The set $A \cap B$ will then be the *empty set*, that is a collection of objects which has no objects in it. The empty set is usually denoted ϕ. Let $a \in U$, then it is true that $a \notin \phi$. We may define ϕ to be $\{x \in U | P(x) \wedge \sim P(x)\}$ using any convenient predicate $P(x)$.

The Difference of Two Sets. Given two sets A and B, we can define a set consisting of all the objects in A which are not in B. This set is denoted $A \setminus B$.

More precisely, let A be defined by $\{x \in U | P(x)\}$ and B by $\{x \in U | Q(x)\}$. Then $A \setminus B$ is defined to be $\{x \in U | P(x) \wedge (\sim Q(x))\}$.

Thus if $a \in A \setminus B$ then $P(a) \wedge (\sim Q(a))$ is true, whence $P(a)$ is true but $Q(a)$ is false. Thus $a \in A$ but $a \notin B$. The following three statements are equivalent.

(i) $a \in A \setminus B$,

(ii) $P(a) \wedge (\sim Q(a))$,

(iii) $(a \in A) \wedge (a \notin B)$.

We illustrate both $A \setminus B$ and the symmetrically defined set $B \setminus A$.

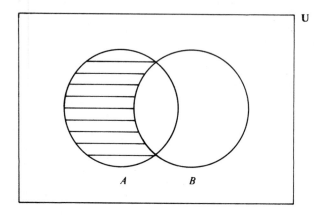

The shaded area represents $A \setminus B$.

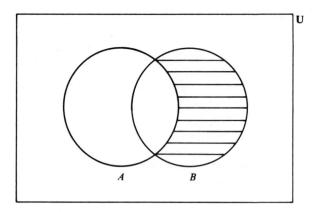

The shaded area represents $B \setminus A$.

The Cartesian Product of Two Sets

Let $P(x)$ and $Q(x)$ be predicates and suppose for the sake of argument that each has just one space. Let us remove the space marker x and leave the space empty. A natural notation for the result is $P(\ \ )$ and $Q(\ \ )$. Next, let us form the predicate $P(\ \ ) \wedge Q(\ \ )$ in which there are two spaces. We must fill the spaces with objects from the universal set **U**.

Now, we have considered the collection of all the statements we can make from $P(\ \ ) \wedge Q(\ \ )$ by taking an object from **U** and putting it in both the spaces in $P(\ \ ) \wedge Q(\ \ )$. If however we put one object from **U** in the space in $P(\ \ )$ and a different object in the space in $Q(\ \ )$, the result will still be a statement. Thus we can consider the collection of statements obtainable from $P(\ \ ) \wedge Q(\ \ )$ by taking a pair of objects from **U** and putting one in each space.

Of course there may in fact be several spaces in $P(\ \ )$ and several in $Q(\ \ )$ and to differentiate between the two types of space occuring in $P(\ \ ) \wedge Q(\ \ )$ we mark the first type with one variable and the second with another, hence $P(x) \wedge Q(y)$. We call a predicate in two variables a *binary predicate*.

Examples of binary predicates are "$(x^2 = 5) \wedge (y = y^2)$"; "$y = 3x + 1$"; "$x + y = y + x$" and "$xy = yx$".

With the predicate $P(x)$, we asked what objects in **U** make a true statement of $P(x)$. For a binary predicate such as $P(x) \wedge Q(y)$, we must ask what *pairs* of objects from **U** make a true statement of $P(x) \wedge Q(y)$.

Thus if we take **R** as our universal set and "$y = 3x + 1$" as our predicate then the pair 0, 1 where 0 replaces x and 1 replaces y results in the true statement "$1 = 3.0 + 1$". It is not sufficient merely to state the pair itself; it must also be clear which object of the pair is to replace which variable. If we replace y by 0 and x by 1 we obtain the false statement "$0 = 3.1 + 1$".

In general, a predicate in two variables is denoted $P(x, y)$. It is convenient to describe x as the first variable and y as the second. If we substitute a pair of objects a and b for x and y respectively then we speak of substituting the *ordered* pair (a, b) where it is to be understood that the first object, a, is to be substituted for the first variable, x. Thus substituting $(0, 1)$ in "$y = 3x + 1$" results in a true statement, while substituting $(1, 0)$ results in a false statement.

The *cartesian product* of a pair of sets A and B is the set of all ordered pairs obtainable by taking an element of A for the first member of the pair and an element of B for the second; it is denoted $A \times B$.

More precisely, if A is defined by $\{x \in U | P(x)\}$ and B by $\{x \in U | Q(x)\}$ then $A \times B$ is defined to be $\{(x, y) | (x \in U) \wedge (y \in U) \wedge P(x) \wedge Q(y)\}$. Here we have two variables each of which will have, in general, its own substitution set although we are using U as the substitution set of all the variables. The ordered pair may be regarded as a single object, compounded from other objects. We make an ordered pair from a and b by enclosing the objects, in the appropriate order, by brackets and separating them by a comma. Thus we can make either (a, b) or (b, a). Since we have constructed a new sort of object using the objects in U, it follows that these new objects are not to be found in the original universal set U. Indeed we should find ourselves in logical difficulties if we assumed they were. Of course we could extend U to include the new objects by forming a new universal set U' which contained U together with the new objects; we should have to alter $P(x)$ to a new predicate "$x \in U \wedge P(x)$" and $Q(x)$ to "$x \in U \wedge Q(x)$". We shall not extend U, this would be turning a convenience into a liability. Notice that we could now use a variable z which is intended to be replaced not by objects in U but by objects from $A \times B$. Thus we can define a set in the form $\{z \in A \times B | P(z)\}$, where the substitution set of z is $A \times B$ and $P(z)$ is a predicate whose spaces are intended to be filled with ordered pairs.

It is not usual in mathematics to bother to create a universal set at all but simply to attach the relevant substitution set to each variable. We are using the universal set for convenience, since we can regard all the variables as having the universal set as their substitution set. This is a convenience to us because we are not interested in any particular substitution set at the moment, whereas in other discussions the substitution sets are important. Notice also that we have changed the space marker in $Q(\ )$ from x to y. Of course we obtain the same statements by substituting elements of U for x in $Q(x)$ as we do by substituting elements of U for y in $Q(y)$, so $\{x \in U | Q(x)\}$ is the same as $\{y \in U | Q(y)\}$.

The following statements are equivalent for $a, b \in U$.

  (i) $(a, b) \in A \times B$
 (ii) $P(a) \wedge Q(b)$
(iii) $(a \in A) \wedge (b \in B)$.

We can draw a diagram to represent this situation.

The lowest horizontal line is intended to represent the objects in the set A laid out in a line while the left-hand vertical line does the same service for B. If we select objects a from A and b from B, draw a

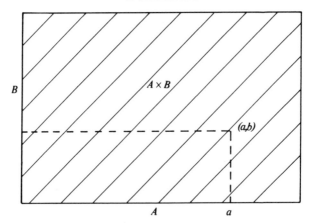

vertical line up from a and a horizontal line across from b then the point at which these lines intersect represents (a, b). The connection with co-ordinate geometry is clear.

Exercises 6

1. Let A be $\{x \in \mathbf{N} \mid x \text{ is even}\}$ and B be $\{x \in \mathbf{N} \mid x \text{ is prime}\}$. Describe the sets $A \cap B$, A', $B \setminus A$.

2. Describe the set of all prime natural numbers without assuming the definition of prime. See page 31.

3. Describe the sets $\{x \in \mathbf{R} \mid (x - 1)(x + 2) = 0\}$ and $\{x \in \mathbf{R} \mid (x + 1)^2 = x^2 + 2x + 1\}$ in a more direct way.

4. Let L be $\{(x, y) \in \mathbf{R} \times \mathbf{R} \mid 3y + 2x + 4 = 0\}$ and M be $\{(x, y) \in \mathbf{R} \times \mathbf{R} \mid y = x + 2\}$.

   Describe $L \cap M$. Using the cartesian plane to represent $\mathbf{R} \times \mathbf{R}$, draw representations of L and M in the plane. Draw also a representation of the set M^{-1} defined to be $\{(y, x) \in \mathbf{R} \times \mathbf{R} \mid y = x + 2\}$.

5. Let A be defined by $\{x \in \mathbf{U} \mid P(x)\}$. Describe in words the elements of the set B defined by $\{P(x) \mid x \in A\}$.

6. Comment on the set A defined to be
   (i) $\{x \in \mathbf{U} \mid x \in A\}$
   (ii) $\{x \in \mathbf{U} \mid x \notin A\}$.

7. Comment on the following introduction of the parameter b. Let b be a natural number which is less than or equal to every natural number.

Answers 6

1. $A \cap B$ is $\{2\}$; A' is $\{x \in \mathbf{N} \mid x \text{ is odd}\}$; $B \setminus A$ is $\{x \in \mathbf{N} \mid x \text{ is a prime number larger than 2}\}$.

2. $\left\{n \in \mathbf{N} \,\middle|\, (n \neq 1) \wedge \left(\frac{n}{2} \notin \mathbf{N}\right) \wedge \left(\frac{n}{3} \notin \mathbf{N}\right) \wedge \left(\frac{n}{4} \notin \mathbf{N}\right) \wedge \ldots \wedge \left(\frac{n}{n-1} \notin \mathbf{N}\right)\right\}.$

3. $\{1, -2\}$ and $\mathbf{R}$.

4. $L \cap M$ is $(-2, 0)$.

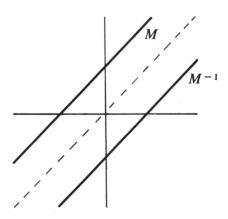

5. The elements of this set are statements. If A were $\mathbf{N}$ then these elements would be $P(1)$, $P(2)$, $P(3)$, ... Further, all the statements in this set of statements are true since an object a only gets into the set A if $P(a)$ is true. The elements of B are not in the original universal set U; if they were we should be in the dangerous situation of having statements which say something about themselves. Consider the old example "this statement is false".

6. (i) This defines A in terms of A and hence does not define A at all. It tells us that $a \in A$ is true if and only if the result of substituting a for x in "$x \in A$" is true. Most helpful. It is however consistent.

(ii) This definition also proposes to define A in terms of A. Not only does it fail to define A, but also no amount of extra information about A is going to help us to define such a set. For this set A has the property that if $a \in A$ then the defining predicate is true so $a \notin A$. On the other hand, if $a \notin A$ then from the defining predicate, $a \notin A$, is false whence $a \in A$. This definition is inconsistent.

Contrasting (i) with (ii), the first definition does not hinder us in our search for a definition of A. Such "definitions", while they do not define things, may constitute a partial definition by placing restrictions on the nature of the thing. Thus (i) tells us at least that A is a set. On the other hand, (ii) is an unmitigated disaster.

7. If we regard this introduction as defining an object b, then we have problems because b is defined in terms of the set of all natural numbers, whilst b itself is an element of this set, and hence we are defining b indirectly in terms of itself. On the other hand, we need not consider ourselves to be defining an object b but rather to be giving the name b to an object which is already well defined. In this case b is an alternative name for the number 1.

CHAPTER 5

QUANTIFICATION

§1 Quantifiers

Suppose we are given a predicate $P(x)$ together with a substitution set $\mathbf{U}$ for x. We may consider the set of statements B defined by $\{P(x) | x \in \mathbf{U}\}$. There are two important statements that can be made about B. The first is that all the statements of B are true. The second is that at least one of the statements of B is true. We wish to formalize these statements.

In order to formalize the statement "all the statements of B are true" we employ the symbol $\forall$ called the *universal quantifier* and write $(\forall x) P(x)$ which may conveniently be read "for all x, $P(x)$". The formal statement $(\forall x) P(x)$ does not mention the substitution set $\mathbf{U}$ for reasons which will become apparent. For the moment we may rectify the situation by rendering the statement as "$(\forall x)P(x) \wedge x \in \mathbf{U}$" and reading it "for all x in $\mathbf{U}$, $P(x)$". Thus for example if $P(x)$ were "$(x + 1)^2 = x^2 + 2x + 1$" and the substitution set of x were $\mathbf{R}$ we could formalize the statement "$(x + 1)^2 = x^2 + 2x + 1$ for all real numbers" as "$(\forall z)((x + 1)^2 = x^2 + 2x + 1) \wedge x \in \mathbf{R}$".

Next, let us formalize the statement "at least one of the statements of B is true". We employ the symbol $\exists$, called the *existential quantifier*, and write $(\exists x)P(x)$. This may conveniently be read "there exists x such that $P(x)$". Again we may rectify the omission of $\mathbf{U}$, writing $(\exists x)P(x) \wedge x \in \mathbf{U}$ and reading "there exists x in $\mathbf{U}$ such that $P(x)$". Hence $(\exists x)(x^2 + 2x + 1 = 0) \wedge x \in \mathbf{R}$ stands for "there is a real number x such that $x^2 + 2x + 1 = 0$". Of course, we can also make the statement $(\forall x)(x^2 + 2x + 1 = 0)$ and $x \in \mathbf{R}$ even though this statement will be false.

We may extend this process to predicates with more than one variable. Let $P(x, y)$ be the predicate "$(x + y)^2 = x^2 + 2xy + y^2$" and let the substitution set of both x and y be $\mathbf{U}$. Then we may render "for all real numbers x and y, $(x + y)^2 = x^2 + 2xy + y^2$" as "$(\forall x)(\forall y)((x + y)^2 = x^2 + 2xy + y^2) \wedge x \in \mathbf{R} \wedge y \in \mathbf{R}$".

Let $Q(x, y)$ be "$x \leqslant y$". Then we can render the statement "for all x there exists y such that $x \leqslant y$, where x and y are real numbers" as $(\forall x)(\exists y)Q(x, y) \wedge x \in \mathbf{R} \wedge y \in \mathbf{R}$. Of course y depends upon x. Using the same predicate we can formalize "There is a first natural number" as $(\exists x)(\forall y)Q(x, y) \wedge x \in \mathbf{R} \wedge \in \mathbf{R}$.

Let $R(x, y)$ be "$2y + x = 1$" then the statement "there exist real numbers x and y such that $2y + x = 1$" can be written $(\exists x)(\exists y)$ $R(x, y) \wedge x \in \mathbf{R} \wedge y \in \mathbf{R}$. Let $P(x, y, z)$ be the predicate "$x + y = z$". Then we can formalize "for all real numbers x and y, $x + y = y + x$" as $(\forall x)(\forall y)(\forall z)(P(x, y, z) \Leftrightarrow P(y, x, z)) \wedge x \in \mathbf{R} \wedge y \in \mathbf{R} \wedge z \in \mathbf{R}$.

This last example demonstrates that formalizing a statement may be no easy task. We observe that "$x + y$" is not a predicate since it does not become a statement, either true or false, when numbers are substituted for x and y. In order to formalize this statement we have had to call to mind the fact that the sum of two real numbers is again a real number. Indeed, part of the use of formalizing statements is that the process draws out hidden or dormant assumptions.

Using several quantifiers raises several questions which we now answer. Let $P(x, y)$ be any predicate in two variables. Is $(\forall x)(\forall y)$ $P(x, y)$ the same as $(\forall y)(\forall x)P(x, y)$? We may consider whether "for all all x and for all y, $P(x, y)$" is the same as "for all y and for all x, $P(x, y)$". Indeed, both statements are the same. If A is the substitution set of x and B the substitution set of y then both statements amount to saying that all the statements in the set $\{P(x, y) | x \in A \cap y \in B\}$ are true. We observe that $(x \in A) \wedge (y \in B) \Leftrightarrow (y \in B) \wedge (x \in A)$ is a tautology, verifiable by truth tables if we take $(x \in A)$ as one statement and $(y \in B)$ as the other. Thus the sets $\{P(x, y) | x \in A \wedge y \in B\}$ and $\{P(x, y) | y \in B \wedge x \in A\}$ are defined by equivalent predicates. We may now wonder whether $(\exists x)(\exists y)P(x, y)$ and $(\exists y)(\exists x)P(x, y)$ are the same. The first claims that one of the statements in the set $\{P(x, y) | x \in A \wedge y \in B\}$ is true and the second that one of the statements in $\{P(x, y) | y \in B \wedge x \in A\}$ is true. Hence again both statements mean the same.

To generalise, it follows that where two quantifiers of the same type lie side by side, they may be interchanged.

Let us now ask whether $(\exists x)(\forall y)P(x, y)$ is the same as $(\forall y)(\exists x)$ $P(x, y)$. We give an example to show that this is not the case. Let $P(x, y)$ be the predicate "$x \geqslant y$" and let $\mathbf{N}$ be the substitution set of both x and y. Then $(\exists x)(\forall y)P(x, y)$ means that there is a natural number which is greater or equal to every natural number, i.e. that there is a greatest natural number. This is false, for let n be any natural number. Then $n + 1$ is a natural number greater than n. On the other hand, $(\forall y)(\exists x)P(x, y)$ means that for each natural number y there is a natural number x greater or equal to y. This is true, since any natural number is greater than or equal to itself.

Statements beginning $(\exists x)(\forall y)$ mean that there is x which will do for all y, whereas statements beginning $(\forall y)(\exists x)$ mean that for each y there is a suitable x but there may be no single x which will do for all y. We may sum the situation up by saying that, irrespective of the nature of $P(x, y)$ and of the substitution sets for x and y it is true that

$$(\exists x)(\forall y)P(x, y) \Rightarrow (\forall y)(\exists x)P(x, y),$$

i.e. if there is an x which will do for all y then for every y there is an x which will do. On the other hand, the statement

$$(\forall y)(\exists x)P(x, y) \Rightarrow (\exists x)(\forall y)P(x, y)$$

is not always true, although of course for certain special $P(x, y)$ and certain substitution sets it may happen to be true; i.e it is not true that because for each y there is an x that will do, there must necessarily be an x which will do for all y.

Let us turn to another question. Let $P(x, y)$ be a binary predicate and let A and B be the substitution sets for x and y respectively. What is the status of $(\forall x)P(x, y)$? Here we have a predicate in two variables, only one of which has been quantified. Let A and B be **R** and let $P(x, y)$ be "$2y + x = 3$". We might ask whether $(\forall x)P(x, y)$ is true. If this is so, then replace x by 1. The result should be a true statement. But the result is "$2y + 1 = 3$" and the truth of this clearly depends on y. In fact, "$2y + 1 = 3$" is a predicate in the single variable y. In general then, $(\forall x)P(x, y)$ becomes a statement if and only if we substitute an object for y from the substitution set of y and it is therefore a predicate in the single variable y. Similarly $(\exists x)P(x, y)$ is a predicate in y, but different from $(\forall x)P(x, y)$. Taking our example "$2y + x = 2$" the predicate $(\exists x)P(x, y)$ becomes a true statement if we substitute 1 for y but $(\forall x)P(x, y)$ becomes a false statement.

We can also make from $P(x, y)$ two different predicates in the single variable x, namely $(\forall y)P(x, y)$ and $(\exists y)P(x, y)$. .

Next, let $k \in A$. What then is the status of $P(k, y)$ where we have substituted the object k for x. Actually, we have answered this question for when we substituted 1 for x in "$2y + x = 3$" the result was "$2y + 1 = 3$" which is clearly a predicate in the single variable y. In general then, replacing x in $P(x, y)$ by an object k from the substitution set of x results in a predicate $P(k, y)$ in the single variable y.

Let us summarize. Given a predicate in the full glory of all its variables, we may take one of two actions in respect of each variable. Either we substitute an object for the variable from its substitution set, or we quantify the predicate with respect to the variable. If we take one of these two courses of action for a given variable, the result is a predicate in the remaining variables. If we continue the procedure until we have taken one of these two courses of action in respect of each variable in the predicate then we obtain a statement.

Let $P(x, y)$ mean "$x \leqslant y$" where $\mathbf{N}$ is the substitution set of each variable. Then $(\exists x)(\forall y)P(x, y)$ means that there is a smallest natural number; $(\forall y)P(1, y)$ says that 1 is the smallest natural number; $(\exists x)P(x, 2)$ says that there is a natural number less than or equal to 2; $P(1, 2)$ says that $1 \leqslant 2$; $P(x, 3)$ is a predicate "$x \leqslant 3$"; $(\exists x)P(x, y)$ is a predicate "There exists a natural number less than or equal to y".

Exercises 7

1. By suitable choice of predicates and substitution sets, formalize the following statements.

   (i) There is no real number x such that $x^2 + 1 = 0$.
   (ii) There exists a real number x and a real number y such that $x^2 + y = 0$.
   (iii) For each real number x, there exists a real number y such that $x^2 + y = 0$.
   (iv) There is no least real number.
   (v) There is no greatest real number.
   (vi) The square of a real number is never negative.
   (vii) If $x = y$ and $y = z$ then $x = z$.
   (viii) There exists a real number l such that for every positive real number ε there exists a positive real number δ such that for all real numbers x, if $-\delta < x - 1 < \delta$ then $-\varepsilon < x^2 + x - l < \varepsilon$.
   (ix) There exists a real number l such that for every positive real number ε there exists a natural number N such that for all natural numbers $n \geqslant N$, $-\varepsilon < \dfrac{n}{2n + 1} - l < \varepsilon$.
   (x) There exists exactly one real number x such that $2x + 1 = 3$.
   (xi) For each integer x there exists an integer y such that $x + y = 0$.
   (xii) For each non-zero real number x, there exists a real number y such that $xy = 1$.

2. What are the principle properties of the concept of order? Consider

the orderings $<$ and $\leqslant$ of integers. Express these properties formally.

3. Let $P(x, y, z)$ mean $xy = z$ and let $S(x, y, z)$ mean $x + y = z$. Express the predicate "$2x + 1 = 0$" in terms of the predicates P and S, giving suitable substitution sets to the variables.

Answers 7

1. (i) Let $P(x)$ be "$x^2 + 1 = 0$". The statement is then $\sim(\exists x)$ $P(x)$ with $x \in \mathbf{R}$, or $(\forall x)(\sim P(x))$ with $x \in \mathbf{R}$.
   (ii) Let $P(x, y)$ be "$x^2 + y = 0$". Then we have $(\exists x)(\exists y)P(x, y)$ where $x \in \mathbf{R}$ and $y \in \mathbf{R}$.
   (iii) Let $P(x, y)$ be "$x^2 + y = 0$". Then we have $(\forall x)(\exists y)P(x, y)$ where $x \in \mathbf{R}$ and $y \in \mathbf{R}$.
   (iv) Let $P(x, y)$ be "$x \leqslant y$". Then we have $\sim(\exists x)(\forall y)P(x, y)$ where $x \in \mathbf{R}$ and $y \in \mathbf{R}$. An alternative is $(\forall x)(\exists y)(\sim P(x, y))$.
   (v) Let $P(x, y)$ be "$x \leqslant y$". Then we have $\sim(\exists y)(\forall x)P(x, y)$ or $(\forall y)(\exists x)(\sim P(x, y))$.
   (vi) Let $P(x)$ be "$x^2 \geqslant 0$". Then we have $(\forall x)P(x) \wedge x \in \mathbf{R}$. This is rather a specialized predicate however. Let $P(x, y, z)$ mean $xy = z$ and let $Q(x, y)$ mean $x \leqslant y$. These are predicates of more general application.

   Then we can define $P(x)$ as $(\forall z)(P(x, x, z) \Rightarrow Q(0, z))$. To show these two predicates are equivalent is to show that for each $k \in \mathbf{R}$, $P(k)$, and $(\forall z)(P(k, k, z) \Rightarrow Q(0, z))$ both have the same truth value.

   (vii) Let $E(x, y)$ mean "$x = y$". Then we have $(\forall x)(\forall y)(\forall z)$ $(E(x, y) \wedge E(y, z) \Rightarrow E(x, z))$. We did not mention a substitution set for x, y and z. Indeed there are many choices, for equality is well-defined on nearly every set of objects.

   (viii) Let $L(l, \varepsilon, \delta, x)$ be $-\varepsilon < x - 1 < \varepsilon \Rightarrow -\delta < x^2 + x - l < \delta$. Then we have $(\exists l)(\forall \varepsilon)(\exists \delta)(\forall x)(L(l, \varepsilon, \delta, x))$.

   Let $\mathbf{R}^+$ denote the set of positive real numbers, then $l \varepsilon \mathbf{R} \wedge \varepsilon \in \mathbf{R}^+ \wedge \delta \in \mathbf{R}^+ \wedge x \in \mathbf{R}$. The predicate $L(l, \varepsilon, \delta, x)$ could be defined in terms of the more basic predicates "$x + y = z$", "$xy = z$" and "$x < y$". Denoting these predicates by $S(x, y, z)$, $P(x, y, z)$ and $Q(x, y)$ the part "$-\varepsilon < x - 1$" becomes, for example, $(\exists y)(\exists z)(S(y, \varepsilon, 0) \wedge S(x, -1, z) \wedge Q(y, z))$. Here, $S(y, \varepsilon, 0)$ effectively introduces y as $-\varepsilon$ while $S(x, -1, z)$ introduces z as $x - 1$. Thus the "sum" predicate can be used to provide subtraction.

(ix) Let $L(l, \varepsilon, N, n)$ be $n \geqslant N \Rightarrow -\varepsilon < \dfrac{n}{2n + 1} - l < \varepsilon$.

Then we have $(\exists l)(\forall \varepsilon)(\exists N)(\forall n)L(l, \varepsilon, N, n)$; where $l \in \mathbf{R}$, $\varepsilon \in \mathbf{R}^+$, $N \in \mathbf{N}$ and $n \in \mathbf{N}$.

(x) Let $E(x, y)$ be "$x = y$" and $P(x)$ be "$2x + 1 = 3$". Then we have $(\exists x)P(x) \wedge (\forall x)(\forall y)(P(x) \wedge P(y) \Rightarrow E(x, y))$, where $x \in \mathbf{R}$ and $y \in \mathbf{R}$. Notice that we must define what is meant by "the same", i.e. equal. Thus $(\forall x)(\forall y)(P(x) \wedge P(y) \Rightarrow E(x, y))$ says that if two objects make $P(x)$ true, then they must be equal. This brings up the point that it is not sufficient to define objects and then assume that two objects are equal if the are "the same". The problem is that we deal in names rather than in objects; the same name can be put in many places at once and there may be many names for the same object. Thus for example we may define the positive fractions as the contents of the set $\left\{ \dfrac{n}{m} \,\middle|\, n \in \mathbf{N} \text{ and } m \in \mathbf{N} \right\}$ but $\frac{2}{2}$ and $\frac{4}{4}$ both are elements of this set. Thus we must define equality on the set if we wish $\frac{2}{2}$ to be the same as $\frac{4}{4}$. In fact we may define $a/b = c/d$ if $ad = bc$ thus defining equality among positive fractions in terms of equality between natural numbers.

(xi) Let $P(x, y)$ be the predicate "$x + y = 0$". Then we have

$$(\forall x)(\exists y)P(x, y) \wedge x \in \mathbf{Z} \wedge y \in \mathbf{Z}.$$

Alternatively we may use the predicate $S(x, y, z)$, meaning $x + y = z$, and write

$$(\forall x)(\exists y)S(x, y, 0) \text{ where } x \in \mathbf{Z} \text{ and } y \in \mathbf{Z}.$$

(xii) Let $P(x, y, z)$ mean $xy = z$.

We may write $(\forall x)(\exists y)P(x, y, 1)$ where the substitution set for x is $\mathbf{R} \setminus \{0\}$ and that of y is $\mathbf{R}$. If this appears a little forced, we may take $E(x, y)$, meaning $x = y$, and write

$$(\forall x)(E(x, 0) \vee (\exists y)P(x, y, 1)) \wedge x \in \mathbf{R} \wedge y \in \mathbf{R}.$$

2. The principal property of ordering, whether we be ordering molecules in a crystal or lengths of wood in a timber yard, is that if one object comes before another and that other before a third then the first object comes before the third. Let $Q(x, y)$ be a binary predicate. Then we express this property

$$(\forall x)(\forall y)(\forall z)(Q(x, y) \wedge Q(y, z) \Rightarrow Q(x, z)).$$

We have given neither a meaning to Q nor substitution sets for the variables. We maintain however that if a meaning is given to Q, and a substitution set provided for x, y and z in such a way that the above statement is true, then the nature of Q will be that it orders the objects in the substitution set.

Let $\mathbf{Z}$ be the substitution set of x, y and z. If $Q(x, y)$ means $x < y$ then the following statements are also characteristic of this type of ordering. Let $E(x, y)$ mean $x = y$.

$$(\forall x)(\forall y)(Q(x, y) \Rightarrow \sim Q(y, x))$$
$$(\forall x)(\forall y)(Q(x, y) \Rightarrow \sim E(x, y))$$
$$(\forall x)(\forall y)(E(x, y) \Rightarrow \sim(Q(x, y) \vee Q(y, x)))$$
$$(\forall x)(\forall y)(Q(x, y) \vee Q(y, x) \vee E(x, y)).$$

These statements amount to saying that exactly one $x < y$, $y < x$, $x = y$ is true for any x, y.

If $Q'(x, y)$ means $x \leqslant y$ then we have a slightly different situation.

$$(\forall x)(\forall y)(Q'(x, y) \vee Q'(y, x))$$
$$(\forall x)(\forall y)(Q'(x, y) \wedge Q'(y, x) \Rightarrow E(x, y)).$$

This says that either $x \leqslant y$ or $y \leqslant x$ for every pair x, y and further that if both $x \leqslant y$ and $y \leqslant x$ then $x = y$.

3. $(\forall z)(P(2, x, z) \Rightarrow S(z, 1, 0)) \wedge x \in \mathbf{R} \wedge \in \mathbf{R}$.

§2 Proof of Quantified Statements

Suppose we are given a predicate $P(x)$ together with a substitution set $\mathbf{U}$ for x. How may we prove $(\forall x)P(x)$?

Now, $(\forall x)P(x)$ is true if and only if all the statements in the set $\{P(x) | x \in \mathbf{U}\}$ are true. If $\mathbf{U}$ is finite, say $\{1, 2\ 3\}$, then it is possible to prove $P(1)$, $P(2)$ and $P(3)$ and this will constitute a proof of $(\forall x)P(x)$. If $\mathbf{U}$ is infinite then it will not be possible to prove the statements in $\{P(x) | x \in \mathbf{U}\}$ separately.

An alternative is to take a typical statement from $\{P(x) | x \in \mathbf{U}\}$ in such a way that we do not know which statement we have taken, and then provide a proof of that statement. If this is possible, then the proof we have found cannot depend on which particular statement of $\{P(x) | x \in \mathbf{U}\}$ we began with and we may expect the proof to work for any statement in $\{P(x) | x \in \mathbf{U}\}$. The proof of the typical statement is held to be a proof of $(\forall x)P(x)$.

The method depends on being deliberately vague about which statement we are proving. This is accomplished by introducing a parameter

k with the incomplete description "let $k \in \mathbf{U}$". We then proceed to prove $P(k)$.

EXAMPLE. Prove that for all natural numbers n, $1 + 2^2 + 2^3 + 2^4 + \ldots + 2^n = \dfrac{1 - 2^{n+1}}{1 - 2}$.

Let $k \in \mathbf{U}$.

With no further information about k we must prove that

$$1 + 2 + 2^2 + 2^3 + \ldots + 2^k = \frac{1 - 2^{k-1}}{1 - 2}.$$

Let S be the number $1 + 2 + 2^2 + \ldots + 2^k$. Here S is a completely described parameter which we use as a short name for $1 + 2 + 2^2 + \ldots + 2^k$. Then we have $S = 1 + 2 + 2^2 + \ldots + 2^{k-1} + 2^k$. Using well known properties of equality between natural numbers we obtain

$$2S = 2 + 2^2 + \ldots + 2^k + 2^{k+1}$$

and

$$S - 2S = 1 - 2^{k+1},$$

whence

$$S = \frac{1 - 2^{k+1}}{1 - 2}.$$

Thus we have deduced the statement

$$1 + 2 + 2^2 + \ldots + 2^k = \frac{1 - 2^{k+1}}{1 - 2}.$$

Since $k \in \mathbf{U}$ constitutes our total knowledge of k we accept as proved

$$(\forall n)\left(1 + 2 + 2^2 + \ldots + 2^n = \frac{1 - 2^{n+1}}{1 - 2}\right)$$

where $\mathbf{N}$ is the substitution set of n.

Next, suppose that we wish to prove a statement of the form $(\exists x) P(x)$. A *constructive proof* of $(\exists x)P(x)$ is one which actually describes an object k and then proves $P(k)$. It is not necessary to justify the procedure by which k was obtained, even though obtaining k is often the major part of the work involved in producing such a proof. A lucky guess is as good as anything. Having obtained an object k however one must prove $P(k)$ in order to show that the k described suitably demonstrates the truth of $(\exists x)P(x)$.

A *non-constructive proof* of $(\exists x)(P(x)$ is actually a proof by contradiction. This we begin by assuming that the statements in the set $\{P(x) \,|\, x \in \mathbf{U}\}$ are all false. It follows from this assumption that the statements in the set $\{\sim P(x) \,|\, x \in \mathbf{U})$ are all true, and hence that $(\forall x)(\sim P(x))$ is true. The proof continues from this point until a contradiction is obtained, see Chapter 3 §4.

The method of contradiction can also be used to prove $(\forall x)(P(x)$. If $(\forall x)P(x)$ is false then not all the statements in $\{P(x) \,|\, x \in \mathbf{U}\}$ are true and hence at least one of the statements in $\{\sim P(x) \,|\, x \in \mathbf{U}\}$ is true. Hence it follows that $(\exists x)(\sim P(x))$ is true. Thus we may introduce a parameter k by "Let k be such that $k \in \mathbf{U}$ and $\sim(P(k))$ is true". The proof develops from this point until a contradiction is obtained.

It often happens that one is interested in finding out whether $(\forall x)P(x)$ is true or false, having no clear idea of which is likely. In order to show this statement true, one would have to provide a proof of $P(k)$ for an arbitrary element k of $\mathbf{U}$. A method of proving it false is to construct a *counter example*, that is to find an object $l \in \mathbf{U}$ such that $P(l)$ is false. This latter method constitutes in fact a proof of $(\exists x)(\sim P(x))$. The initial problem is whether one should try for a proof of $P(k)$ or whether to look for a counter example. Whichever path one takes, one is liable to expose the behaviour of $P(x)$; it may easily happen that an attempt to produce a proof leads to a method of constructing a counter example or vice versa.

We now make a remark on what can be deduced *from* quantified statements. Let $(\forall x)\,P(x)$ be true, where the substitution set of x is $\mathbf{U}$. Then for any object k in $\mathbf{U}$ we may deduce that $P(k)$ is true. This is linked to the method of proof of $(\forall x)P(x)$, which constituted in effect a method of proving $P(k)$ suitable for any k in $\mathbf{U}$. Let $(\forall x)P(x)$ and $(\forall x)Q(x)$ both be true. Let $k \in \mathbf{U}$, then $P(k)$ and $Q(k)$ are true, whence $P(k) \wedge Q(k)$ is true. This constitutes a proof that $(\forall x)P(x) \wedge (\forall x)Q(x) \Rightarrow (\forall x)(\mathbf{P}(x) \wedge Q(x))$.

Next, let $(\exists x)P(x)$ be true. Then we may introduce a parameter k by "Let k be such that $k \in \mathbf{U}$ and $P(k)$ is true". It follows now that $P(k)$ is true. It is essential that k shall not already have been introduced, since we would then be using k as the name of two possibly different objects.

Let $(\exists x)P(x)$ and $(\exists x)Q(x)$ both be true. We may introduce k by "Let k be such that $k \in \mathbf{U}$ and $P(k)$ is true" and deduce $P(k)$. We may also introduce l by "Let l be such that $l \in \mathbf{U}$ and $Q(l)$ is true". It follows that $Q(l)$ and we may also deduce $P(k) \wedge Q(l)$ but this does not prove $(\exists x)(P(x) \wedge Q(x))$. Thus there exists a first natural number and there exists an even natural number, but there does not exist a natural number which is both a first natural number and an even natural number.

We conclude this section with an example.

Prove that there exists a real number l such that for all positive real numbers ε there exists a positive real number δ such that for all real numbers x if $-\delta < x - 1 < \delta$ then $-\varepsilon < (4x + 1) - l < \varepsilon$.

We may analyse this statement to obtain $(\exists l)(\forall \varepsilon)(\exists \delta)(\forall x)$ $((-\delta < x - 1 < \delta) \Rightarrow (-\varepsilon < (4x + 1) - l < \varepsilon))$. We deal with the quantifiers one at a time. Thus we first see that we wish to prove a statement of the form $(\exists l)P(l)$. Our proof will be constructive. We shall prove $(\exists l)P(l)$ by proving $P(5)$. Thus we have to prove

$$(\forall \varepsilon)(\exists \delta)(\forall x)(-\delta < x - 1 < \delta \Rightarrow -\varepsilon < (4x + 1) - 5 < \varepsilon).$$

Our guess at 5 is based on the intuitive notion that we ought to be able to make $4x + 1$ close to 5 by taking a value of x close to 1. We now have to prove a statement of the form $(\forall \varepsilon) Q(\varepsilon)$. Let $e \in \mathbf{R}^+$. We confine our knowledge of e to the fact that it is an element of the substitution set of ε. We now have to prove $Q(e)$, i.e.

$$(\exists \delta)(\forall x)(-\delta < x - 1 < \delta \Rightarrow -e < (4x + 1) - 5 < e).$$

This is a statement of the form $(\exists \delta)R(\delta)$ of which we give a constructive proof. We shall in fact prove $R(e/4)$. Notice that the status of e as a parameter is that it is the name of a definite number, so that $e/4$ is a definite number. We have now to prove $R(e/4)$, i.e.

$$(\forall x)(-e/4 < x - 1 < e/4 \Rightarrow -e < (4x + 1) - 5 < e).$$

This is a statement of the form $(\forall x)S(x)$.

The substitution of x is $\mathbf{R}$. Therefore, let $k \in \mathbf{R}$. With no additional knowledge of k we have to prove $S(k)$, i.e.

$$-e/4 < k - 1 < e/4 \Rightarrow -e < (4k + 1) - 5 < e.$$

We shall prove this implication by the direct method. Assume that $-e/4 < k - 1 < e/4$ is true. We must prove $-e < (4k + 1) - 5 < e$. We take as a correct rule of inference that if a true inequality be multiplied through by a positive real number, then the resulting inequality is also true. Thus from

$$-e/4 < k - 1 < e/4$$

we deduce

$$-e < 4k - 4 < e.$$

Hence $-e < (4k + 1) - 5 < e$, which was to have been proved. We have used, without reference, some properties of multiplication and addition of real numbers. Now, we have proved $-e < (4k + 1) - 5 < e$ under the assumption

$$-e/4 < k - 1 < e/4$$

so that

$$-e/4 < k - 1 < e/4 \Rightarrow -e < (4k + 1) - 5 < e.$$

Since k could have been any element of $\mathbf{R}$ we conclude

$$(\forall x)(-e/4 < x - 1 < e/4 \Rightarrow -e < (4x + 1) - 5 < e), \; x \in \mathbf{R}.$$

Now $e/4$ is an element of $\mathbf{R}^+$, so

$$(\exists \delta)(\forall x)(-\delta < x - 1 < \delta \Rightarrow -e < (4x + 1) - 5 < e)$$

where $x \in \mathbf{R}$ and $\delta \in \mathbf{R}^+$.

Now e could have been any element of $\mathbf{R}^+$, so

$$(\forall \varepsilon)(\exists \delta)(\forall x)(-\delta < x - 1 < \delta \Rightarrow -\varepsilon < (4x + 1) - 5 < \varepsilon),$$

where $x \in \mathbf{R}$, $\delta \in \mathbf{R}^+$ and $\varepsilon \in \mathbf{R}^+$. Finally, $5 \in \mathbf{R}$ so

$$(\exists l)(\forall \varepsilon)(\exists \delta)(\forall x)(-\delta < x - 1 < \delta \Rightarrow -\varepsilon < (4x + 1) - l < \varepsilon),$$

where $x \in \mathbf{R}$, $\delta \in \mathbf{R}^+$, $\varepsilon \in \mathbf{R}^+$ and $l \in \mathbf{R}$.

In practice of course, such a proof would be presented in much less detail. One potentially dangerous feature of a more normal description of this proof is that the symbol ε would be made to do service not only in its capacity as a variable, but also for the parameter e. Similarly, x would have been used where we introduced the parameter k, as well as where we have used it as a variable.

Exercises 8

Assume the usual properties of multiplication, addition, order and equality for the real numbers. Analyse the following statements and prove them, describing the method of proof and differentiating between letters used as variables and those used as parameters.

    (i) There is no largest real number.

    (ii) There is no smallest positive real number.

    (iii) There is no real number x such that $x^2 + 1 = 0$.

    (iv) If for each real number K there exists a natural number N such that $N \geqslant K$, then for every real number R there exists a natural number M such that for all natural numbers $m \geqslant M$, $\frac{1}{2}m - 5 \geqslant R$.

(v) If for each real number K there exists a natural number N such that $N \geqslant K$, then for every positive real number x and for every positive real number y there exists a natural number n such that $y/n < x$.

Answers 8

(i) A suitable analysis would be $\sim(\exists x)(\forall y)(y \leqslant x)$ or, equivalently, $(\forall x)(\exists y)(y > x)$, where $\mathbf{R}$ is the substitution set of both x and y. We use the latter formalization. Let k be an arbitrary element of $\mathbf{R}$. We have to prove $(\exists y)(y > k)$. We use the constructive method; a suitable object is $(k + 1)$. We are left to prove that $k + 1 > k$, which is a consequence of the properties of addition and order.

(ii) A suitable analysis would be $\sim(\exists x)(\forall y)(x \leqslant y)$ or $(\forall x)(\exists y)$ $(y < x)$ where the substitution set of both x and y is $\mathbf{R}^+$. We use the latter formalisation. Let k be an arbitrary element of $\mathbf{R}^+$. We now have to prove $(\exists y)(y < k)$. A suitable choice of object for a constructive proof is $k/2$. We note that since $k \in \mathbf{R}^+$ it follows that $k/2 \in \mathbf{R}^+$. Now we have to prove that $k/2 < k$, which is evidently true.

(iii) There is no real number x such that $x^2 + 1 = 0$. A formalisation might be $\sim(\exists x)(x^2 + 1 = 0)$ or equivalently $(\forall x)$ $(x^2 + 1 \neq 0)$. We prove the latter formulation. Let k be an arbitrary element of $\mathbf{R}$. We have to prove that $k^2 + 1 \neq 0$. We know that $(\forall x)(x^2 \geqslant 0)$ is true where $\mathbf{R}$ is the substitution set of x. Thus all the statements in $\{x^2 \geqslant 0 | x \in \mathbf{R}\}$ are true. In particular $k \in \mathbf{R}$ so $k^2 \geqslant 0$ is true. Hence $k^2 + 1 > 0$ and so $k^2 + 1 \neq 0$.

(iv) $(\forall K)(\exists N)(N \geqslant K) \Rightarrow (\forall R)(\exists M)(\forall m)(m \geqslant M \Rightarrow \frac{1}{2}m - 5 \geqslant R)$. Here $\mathbf{N}$ is the substitution set of N, M and m while $\mathbf{R}$ is the substitution set of K and R. We are faced initially with an implication, which we shall prove by the direct method.

(a) Assume that $(\forall K)(\exists N)(N \geqslant K)$ is true. We have to prove $(\forall R)(\exists M)(\forall m)(m \geqslant M \Rightarrow \frac{1}{2}m - 5 \geqslant R)$. Let r be an arbitrary element of $\mathbf{R}$. We have to prove $(\exists M)(\forall m)(m \geqslant M \Rightarrow \frac{1}{2}m - 5 \geqslant r)$.

We intend to provide a constructive proof so we must find a suitable natural number to substitute for M in $(\forall m)(m \geqslant M \Rightarrow \frac{1}{2}m - 5 \geqslant r)$. Examining $\frac{1}{2}m - 5 \geqslant r$ it rather looks as though $M = 2(r + 5)$ might do, except that $2(r + 5)$ may not be a natural number. We employ (a), which assures us that all the statements in $\{(\exists N)(N \geqslant K)|K \in \mathbf{R}\}$ are true. In particular, $(\exists N)(N \geqslant 2(r + 5))$ is true. Let N_0 be a natural number such that $N_0 \geqslant 2(r + 5)$. This number N_0 is the one we shall substitute for M, yielding $(\forall m)(m \geqslant N_0 \Rightarrow \frac{1}{2}m - 5 \geqslant r)$ as the statement to be proved.

Let p be an arbitrary element of $\mathbf{N}$. We have now to prove $p \geqslant N_0 \Rightarrow \frac{1}{2}p - 5 \geqslant r$. We use the direct method. Assume that $p \geqslant N_0$ is true. Recalling the introduction of N_0, we know that $N_0 \geqslant 2(r + 5)$. Hence $p \geqslant 2(r + 5)$, whence $\frac{1}{2}p \geqslant r + 5$ and finally $\frac{1}{2}p - 5 \geqslant r$ which was to have been proved.

We have used r, N_0 and p as parameters. Notice that the introduction of a parameter shortens the statement to be proved by dispensing with a quantifier.

(v) $(\forall K)(\exists N)(N \geqslant K) \Rightarrow (\forall x)(\forall y)(\exists n)(y/n < x)$. Here $\mathbf{R}$ is the substitution set of K, $\mathbf{R}^+$ of x and y and $\mathbf{N}$ of N and n. We use the direct method for the implication.

(b) Assume $(\forall K)(\exists N)(N \geqslant K)$. We have now to prove $(\forall x)(\forall y)$ $(\exists n)(y/n < x)$. Let a be an arbitrary element of $\mathbf{R}^+$. We must prove $(\forall y)(\exists n)(y/n < a)$. Let b be an arbitrary element of $\mathbf{R}^+$; we must use no information about b other than the fact that it is an element of $\mathbf{R}^+$, in particular we cannot assume that it is the same object as a. We emphasise that a and b are the names of element $\mathbf{R}^+$, but we do not know of which elements of $\mathbf{R}^+$ they are the names; thus in particular it would be wrong to make any assumption which would imply that they were different names for the same "arbitrary element of $\mathbf{R}^+$".

We now have to prove $(\exists n)(b/n < a)$. We use the constructive method. It is evident from $b/n < a$ that any natural number $> b/a$ will do. Of course b/a may not itself be a natural number so we cannot guarantee that $b/a + 1$, which is certainly larger than b/a, is a natural number. We appeal to (b) which informs us that all the statements of $\{(\exists N)(N \geqslant K)|K \in \mathbf{R}\}$ are true. In particular therefore $(\exists N)(N \geqslant (b/a + 1))$ is true. Let N_0 be a natural number such that $N_0 \geqslant (b/a + 1)$. We substitute N_0 for n in $(b/n < a)$ and thus we must prove $b/N_0 < a$.

By the introduction of N_0, $N_0 \geqslant (b/a + 1) > b/a$ whence $aN_0 > b$ and so $b/N_0 < a$ as was to have been proved.

We have used as parameters a, b and N_0.

§3 The Induction Theorem

The purpose of the induction theorem is to give a method of proving statements of the form $(\forall x)P(x)$, when the substitution set of x is a set which is ordered in a manner typified by the usual ordering by magnitude of N.

We observe that the following properties are possessed by the usual ordering $<$ of N.

(i) Any natural number has a unique successor. That is, if x is a natural number then there exists a unique natural number y which is the smallest natural number greater than x. For example, the successor of 2 is 3, of 11 is 12 and of k is $(k + 1)$.

(ii) Any non-empty set of natural numbers contains an element smaller than all the other elements in the set. For example the smallest element of N is 1, the smallest element of the set of all even natural numbers is 2 and the smallest element of the set of all prime factors of 75 is 3.

(iii) Every natural number except the first has a unique predecessor. That is, if x is a natural number and x is not the first natural number, then there exists a unique natural number y which has x as its successor. The predecessor of 2 is 1, of 11 is 10, of k is $(k - 1)$ providing $k \neq 1$.

(iv) The first natural number is the successor of no natural number.

A simple form of the induction theorem may be stated in words as follows.

Let $P(x)$ be a predicate and let N be the substitution set of x. The induction theorem says that the statement $(\forall x)P(x)$ is implied by the two statements (a) $P(x)$ is true for the first element of N and (b) if $P(x)$ should be true for a natural number k, then it is also true for the next natural number after k, i.e. $k + 1$.

Thus instead of proving $(\forall x)P(x)$ directly we may instead prove (a) and (b) and then deduce $(\forall x)P(x)$ quoting the induction theorem. The statement (a) simply claims that $P(1)$ is true and its proof is therefore simply a proof of $P(1)$. The original phrasing of (a) was intended to emphasise that we must be able to interpret the adjective "first" for the objects in N.

The statement (b) does not claim that there is a natural number for which $P(x)$ holds; it claims that *if* $P(x)$ holds for a natural number, then it holds for the next natural number. Of course we must be able to interpret the adjective "next". A proof of (b) is a proof of the statement $P(k) \Rightarrow P(k + 1)$ in which the only information used about k is the fact that it is a natural number.

EXAMPLE. Prove that for all natural numbers n,

$$1 + 2 + 3 + \ldots + (n - 1) + n = \tfrac{1}{2}n(n + 1).$$

The predicate $P(n)$ is "$1 + 2 + 3 + \ldots + (n - 1) + n = \tfrac{1}{2}n$ $(n + 1)$". First we must prove $P(1)$, that is $1 = \tfrac{1}{2} \cdot 1 \cdot (1 + 1)$ which is obviously true. Next let $k \in \mathbf{N}$. We must prove $P(k) \Rightarrow P(k + 1)$. We use the direct method. Assume that $P(k)$ is true. Thus $1 + 2 + 3 + \ldots + k = \tfrac{1}{2}k(k + 1)$. We must deduce $P(k + 1)$. From $P(k)$ it follows that

$$1 + 2 + 3 + \ldots + k + (k + 1) = \tfrac{1}{2}k(k + 1) + (k + 1).$$

We have simple added $(k + 1)$ to both sides of an equation. Rearranging the right hand side,

$$\tfrac{1}{2}k(k + 1) + (k + 1) = (\tfrac{1}{2}k + 1)(k + 1) = \tfrac{1}{2}(k + 1)(k + 2).$$

Hence $1 + 2 + 3 + \ldots + (k + 1) = \tfrac{1}{2}(k + 1)((k + 1) + 1)$. This is actually $P(k + 1)$. Thus we have proved (a) and (b) and may quote the induction theorem in order to deduce $(\forall n)P(n)$.

It is not essential that the substitution set of n be $\mathbf{N}$ in order that we may use the induction theorem in order to prove $(\forall n)P(n)$. What is essential is that we be able to interpret the words "first" and "next" for the substitution set we are using. In particular, we might chop off the beginning of $\mathbf{N}$. In the next example the substitution set is $\{2, 3, 4, \ldots\}$ which can be described as $\{n \in \mathbf{N} \mid n \neq 1\}$, $\{n \in \mathbf{N} \mid n > 1\}$, $\{(n + 1) \mid n \in \mathbf{N}\}$ or $\mathbf{N} \setminus \{1\}$. For this set, the first element is 2 while the next element after k is $k + 1$.

EXAMPLE. Prove that for all natural numbers greater than 1, $n^3 - n$ is divisible by 3. Here $p(n)$ is "$n^3 - n$ is divisible by 3" and the substitution set for n is $\mathbf{N} \setminus \{1\}$.

First we must prove $P(2)$ which says that "$2^3 - 2$ is divisible by 3". This is true because $2^3 - 2 = 8 - 2 = 6 = 2 \times 3$.

Next let $k \in \mathbf{N} \setminus \{1\}$. We must prove $P(k) \Rightarrow P(k + 1)$. We use the direct method. Assume $P(k)$. Thus $k^3 - k$ is divisible by 3. Consider $(k + 1)^3 - (k + 1)$. Now $(k + 1)^3 - (k + 1) = (k^3 + 3k^2 + 3k + 1) - (k + 1) = (k^3 - k) + 3(k^2 + k)$. Since $k^3 - k$ is divisible by 3 it follows that $(k + 1)^3 - (k + 1)$ is divisible by 3; but this statement is just $P(k + 1)$. We provide two justifications for the induction theorem, the first is intuitive.

If we wish to prove $(\forall x)P(x)$ where x has substitution set $\mathbf{N}$, we wish to prove the list of statement $P(1)$, $P(2)$, $P(3)$, ... When we are going to use induction, we prove $P(1)$ together with all statements of the form $P(k) \Rightarrow P(k + 1)$. Thus we prove $P(1)$ and the list of statements $P(1) \Rightarrow P(2)$, $P(2) \Rightarrow P(3)$, $P(3) \Rightarrow P(4)$, ... It seems reasonable that we use $P(1)$ and $P(1) \Rightarrow P(2)$ to deduce $P(2)$; we can now use $P(2)$ with $P(2) \Rightarrow P(3)$ to deduce $P(3)$; then $P(3)$ and $P(3) \Rightarrow P(4)$ yield $P(4)$ and so on. Thus this process ought to allow us to deduce $P(1)$, $P(2)$, $P(3)$, ... successively as required. The trouble with this argument is that although the process of deducing $P(1)$, $P(2)$, $P(3)$, ... from $P(1)$, $P(1) \Rightarrow P(2)$, $P(2) \Rightarrow P(3)$, ... seems as though it proceeds in a satisfactory manner, it is nevertheless an infinite process and we shall never get to the end of it. Thus we can never complete the list of deductions $P(1)$, $P(2)$, $P(3)$, ... and be able to claim $(\forall x)P(x)$. Of course the process is so simple that one might be prepared to "trust" it after one has seen how it operates. Such trust is too often misplaced.

We give a proof of the induction theorem based on the property (ii) of natural numbers. We shall in fact show that ((statement (a)) $\wedge$ (statement (b)) $\Rightarrow (\forall x)P(x)$).

This implication we prove by contradication. Thus we assume statements (a) and (b) are true but that the statement $(\forall x)P(x)$ is false.

Since $(\forall x)P(x)$ is false, the set $\{x \in \mathbf{N} \,|\, {\sim}P(x)\}$ is not empty. By (ii) it has a first element. Let l be the first element of this set. It follows that $P(l)$ is false. Consider $P(l - 1)$. Now l is the first element of $\mathbf{N}$ such that $P(l)$ is false. Hence $P(l - 1)$ is true. We observe that statement (a) informs us that $P(1)$ is true so that since $P(l)$ is false we must have $l \neq 1$. This is important because were $l = 1$ then $(l - 1)$ would not be a natural number and so the substitution of $(l - 1)$ in $P(x)$ would not be justified.

Now, statement (b) informs us that all statements $P(k) \Rightarrow P(k + 1)$ are true, for $k \in \mathbf{N}$. Hence in particular $l - 1 \in \mathbf{N}$ so that $P(l - 1) \Rightarrow P(l)$. Since $P(l - 1)$ is true we deduce that $P(l)$ is true, contradiction since we have already noted that $P(l)$ is false. Hence we have deduced a contradiction from the assumption that the induction theorem is false. This concludes the proof.

Another version of the induction theorem which is an extension of the previous version, is the following.

Let $P(x)$ be a predicate and let $\mathbf{N}$ be the substitution set of x. The induction theorem says that the statement $(\forall x)P(x)$ is implied by the statement (A) "If $P(x)$ is true for all natural numbers smaller than the natural number k, then it is true for k".

In order to prove (A) we must take an arbitrary element $k \in \mathbf{N}$ and prove $P(k)$ on the assumption that $P(x)$ holds for all natural numbers smaller than k.

What does the assumption "$P(x)$ holds for all natural numbers smaller than 1" mean? Since there are no natural numbers smaller than 1, we are not assuming $P(x)$ to hold for any natural number. Hence in proving (A) we have in effect proved $P(1)$ without any assumption. Thus to prove (A) prove $P(1)$ together with all statements of the form

$$P(1) \wedge P(2) \wedge P(3) \wedge \ldots \wedge P(k - 1) \Rightarrow P(k)$$

where k is a natural number greater than 1. Thus we are required to prove the list of statements: $P(1)$, $P(1) \Rightarrow P(2)$, $P(1) \wedge P(2) \Rightarrow P(3)$, $P(1) \wedge P(2) \wedge P(3) \Rightarrow P(4)$, ... The theorem assures us that if we can do this then we may deduce that $P(1)$, $P(2)$, $P(3)$, ... are all true. We can see intuitively that this ought to be so, for from $P(1)$ and $P(1) \Rightarrow P(2)$ we may deduce $P(2)$; now from $P(1) \wedge P(2)$ together with $P(1) \wedge P(2) \Rightarrow P(3)$ we may deduce $P(3)$; thus we now have $P(1) \wedge P(2) \wedge P(3)$ which together with $P(1) \wedge P(2) \wedge P(3) \Rightarrow P(4)$ yields $P(4)$, and so on.

The proof of this form of the induction theorem is given as an exercise.

The induction theorem is a method of proving $(\forall x)P(x)$ which works only when the substitution set of x has sufficient structure for us to determine the meaning of "first" and "next". We shall leave the question of just what is required of a substitution set in order that one may use induction as a method of proof. We observe only that $\mathbf{N}$ together with sets of the form $\{n \in \mathbf{Z} \mid n \geqslant k\}$ where $k \in \mathbf{Z}$ are suitable. Thus $\{0, 1, 2, 3, \ldots\}$ and $\{-1, 0, 1, 2, \ldots\}$ are suitable.

Examples 9

1. Prove, using property (ii), that $(\forall x)P(x)$ is implied by the statement (A) of §3.

2. Prove by induction that for all natural numbers n.

   (i) $1^2 + 2^2 + 3^2 + \ldots + n^2 = \dfrac{1}{6}n(n + 1(2n + 1)$

   (ii) $1^3 + 2^3 + 3^3 + \ldots + n^3 = \dfrac{1}{4}n^2(n + 1)^2$

   (iii) The nth derivative of $\dfrac{1}{1 + x}$ is $(-1)^n n!(1 + x)^{-(n+1)}$ (Here $n!$ denotes $1 \times 2 \times 3 \times \ldots \times n$).

(iv) $\dfrac{1}{1^2} + \dfrac{1}{2^2} + \dfrac{1}{3^2} + \ldots + \dfrac{1}{n^2} < 2 - \dfrac{1}{n}.$

3. Prove by induction that for all natural numbers greater than 3, $n! > 2^n$.

Answers 9

1. By contradiction. Suppose it is true that whenever $P(x)$ is true for all natural numbers less than a certain natural number k then it is true for k, but that $(\forall x)P(x)$ is false. Since $(\forall x)P(x)$ is false, the set $\{x \in \mathbf{N} \mid \sim P(x)\}$ is not empty and by property (ii) contains a first element. Let l be that first element. Then by definition of l, $P(x)$ is true for all natural numbers less than l and hence it is true for l and $P(l)$ is true. But $l \in \{x \in \mathbf{N} \mid \sim P(x)\}$ so $P(l)$ is false. This concludes the proof by contradiction.

2. (i) Firstly $1^2 = \dfrac{1}{6}1(1 + 1)(2 \cdot 1 + 1)$. Secondly, let $k \in \mathbf{N}$ and

assume $1^2 + 2^2 + \ldots + k^2 = \dfrac{1}{6}k(k + 1)(2k + 1)$.

Then

$$1^2 + 2^2 + \ldots + k^2 + (k + 1)^2 = \dfrac{1}{6}k(k + 1)(2k + 1) + (k + 1)^2$$

$$= \dfrac{1}{6}(k + 1)(2k^2 + k + 6(k + 1))$$

$$= \dfrac{1}{6}(k + 1)((k + 1) + 1)(2(k + 1) + 1.$$

Hence

$$1^2 + 2^2 + \ldots + (k + 1)^2$$

$$= \dfrac{1}{6}(k + 1)((k + 1) + 1)(2(k + 1) + 1).$$

(ii) Firstly $1^3 = \dfrac{1}{4}1^2(1 + 1)^2$. Secondly, let $k \in \mathbf{N}$ and assume

$1^3 + 2^3 + \ldots + k^3 = \dfrac{1}{4}k!(k + 1)^2$.

Then

$$1^3 + 2^3 + \ldots + k^3 + (k + 1)^3 = \dfrac{1}{4}k^2(k + 1)^2 + (k + 1)^3$$

$$= \dfrac{1}{4}(k + 1)^2(k^2 + 4(k + 1)) = \dfrac{1}{4}(k + 1)^2((k + 1) + 1)^2.$$

Hence

$$1^3 + 2^3 + \ldots + (k + 1)^3 = \frac{1}{4}(k + 1)^2((k + 1) + 1)^2.$$

(iii) The 1st derivative of $\dfrac{1}{1 + x}$ is $(-1)^1 1!(1 + x)^{-(1+1)}$.

Next, let $k \in \mathbf{N}$. Assume that the kth derivative of $\dfrac{1}{1 + x}$ is

$(-1)^k k!(1 + x)^{-(k+1)}$. Differentiating again, the $(k + 1)^{th}$

derivative of $\dfrac{1}{1 + x}$ is $\dfrac{d}{dx}((-1)^k k!(1 + x)^{-(k+1)})$

$$= (-1)^k . k! . (-(k + 1))(1 + x)^{-(k+1)-1}$$

$$= (-1)^{k+1}(k + 1)! . (1 + x)^{-((k+1)+1)}.$$

Thus the $(k + 1)^{th}$ derivative of $\dfrac{1}{1 + x}$ is

$$(-1)^{k+1} . (k + 1)! . (1 + x)^{-((k+1)+1)}.$$

(iv) Firstly $\dfrac{1}{1^2} < 2 - \frac{1}{2}$. Next, let $k \in \mathbf{N}$. Assume that

$$\frac{1}{1^2} + \frac{1}{2^2} + \frac{1}{3^2} + \ldots + \frac{1}{k^2} < 2 - \frac{1}{k}.$$

Then

$$\frac{1}{1^2} + \frac{1}{2^2} + \frac{1}{3^2} + \ldots + \frac{1}{k^2} + \frac{1}{(k + 1)^2} < 2 - \frac{1}{k} + \frac{1}{(k + 1)^2}$$

$$= 2 - \frac{1}{k} + \frac{1}{k^2 + 2k + 1} < 2 - \frac{1}{k} + \frac{1}{k^2 + k}$$

$$= 2 - \frac{1}{k} + \frac{1}{k} - \frac{1}{k + 1} = 2 - \frac{1}{k + 1}.$$

Hence

$$\frac{1}{1^2} + \frac{1}{2^2} + \ldots + \frac{1}{(k + 1)^2} < 2 - \frac{1}{(k + 1)}.$$

3. Firstly $24 = 4! > 2^4 = 16$. Next, let k be a natural number $\geqslant 4$. Assume $k! > 2^k$. Then $k! . (k + 1) > 2^k . (k + 1)$. Hence since $k + 1 \geqslant 2$, $(k + 1)! > 2^{k+1}$.

§4 Set Equality and Inclusion

We have discussed how sets may be defined and how new sets may be created from old. We have not yet discussed any relationships which may hold between sets. One of the first problems is to decide when two sets are to be *equal*.

Intuitively, sets are collections of objects and two collections may reasonably be regarded as identical if they contain exactly the same objects.

More precisely, let A and B be sets. We define $A = B$ to mean $(\forall x)(z \in A \Leftrightarrow x \in B)$. We may go back one stage further. Let A be the set $\{x \in \mathbf{U} \,|\, P(x)\}$ and let B be $\{x \in \mathbf{U} \,|\, Q(x)\}$. Then $A = B$ is defined to mean $(\forall x)(P(x) \Leftrightarrow Q(x))$. Thus let $k \in \mathbf{U}$. Then $P(k)$ is true if and only if $Q(k)$ is true and so $k \in A$ if and only if $k \in B$.

A proof of the statement $A = B$ consists of a proof of $(\forall x)(x \in A \Leftrightarrow x \in B)$. In order to prove this last statement we first render it as $(\forall x)[(x \in A \Rightarrow x \in B) \wedge (x \in B \Rightarrow x \in A)]$ and finally as $(\forall x)(x \in A \Rightarrow x \in B) \wedge (\forall x)(x \in B \Rightarrow x \in A)$. This makes it clear that we have in fact to prove two statements. We shall have to supply a proof of $(\forall x)(x \in A \Rightarrow x \in B)$ and a proof of $(\forall x)(x \in B \Rightarrow x \in A)$.

EXAMPLE. Let x be a real number. We define $|x|$ such that

(i) $|x| = x$ if x is positive and

(ii) $|x| = -x$ otherwise.

This defines the modulus of a real number x.

Let A be $\{x \in \mathbf{R} \,|\, |x - 3| \leqslant 2\}$ and let B be $\{x \in \mathbf{R} \,|\, 1 \leqslant x \leqslant 5\}$. We prove that $A = B$. Firstly let us prove $(\forall x)(x \in A \Rightarrow x \in B)$. Let $k \in \mathbf{R}$. We prove $k \in A \Rightarrow k \in B$ by the direct method. Let $k \in A$ be true. Then $|k - 3| \leqslant 2$ is true. Thus *either* $k - 3 > 0$ and $|k - 3| = k - 3 \leqslant 2$ *or* $k - 3 < 0$ and $|k - 3| = -(k - 3) \leqslant 2$. Hence *either* $k > 3$ and $k \leqslant 5$ *or* $k \leqslant 3$ and $k \geqslant 1$. Hence either $3 < k \leqslant 5$ or $1 \leqslant k \leqslant 3$. Hence $1 \leqslant k \leqslant 5$. Thus $k \in B$.

Next we must prove $(\forall x)(x \in B \Rightarrow x \in A)$. Let $k \in \mathbf{R}$ and assume $k \in B$. We must deduce that $k \in A$. Since $k \in B$, $1 \leqslant k \leqslant 5$. Thus either $1 \leqslant k \leqslant 3$ or $3 < k \leqslant 5$. Thence $1 - 3 \leqslant k - 3 \leqslant 3 - 3$ or $3 - 3 < k - 3 \leqslant 5 - 3$, i.e. $-2 \leqslant k - 3 \leqslant 0$ or $0 < k - 3 \leqslant 2$. Hence $2 \geqslant -(k - 3) \geqslant 0$ or $0 < k - 3 \leqslant 2$. Hence $|k - 3| \geqslant 2$ and so $k \in A$.

We may now conclude that $A = B$.

EXAMPLE. Prove that $(A \cup B)' = A' \cap B'$ for any pair of sets A and B.

Firstly we prove that $(\forall x)(x \in (A \cup B)' \Rightarrow x \in A' \cap B')$. We suppose there is a universal set $\mathbf{U}$ which supplies objects for A and B and that the complements indicated are with respect to $\mathbf{U}$. Thus the substitution set for x is $\mathbf{U}$. Let $k \in \mathbf{U}$. Assume $k \in (A \cup B)'$. Then, from the definition of complement, $k \notin A \cup B$. From the definition of union it follows that $k \notin A$ and $k \notin B$. But if $k \notin A$ then $k \in A'$ and if $k \notin B$ then $k \in B'$. Hence k is an element of both A' and B' and therefore, from the definition of intersection, $k \in A' \cap B'$ as required. Next, we must prove $(\forall x)(x \in A' \cap B' \Rightarrow x \in (A \cup B)')$. Let $k \in \mathbf{U}$. Assume $k \in A' \cap B'$. Then $k \in A'$ and $k \in B'$. Hence $k \notin A$ and $k \notin B$ so that $k \notin A \cup B$ which means that $k \in (A \cup B)'$ as required.

Thus we conclude that $(A \cup B)' = A' \cap B'$; this equality expresses something about the nature of $\cup$, $\cap$ and $'$. Let $\mathscr{C}$ be a set whose elements are themselves sets. We have in fact proved that $(\forall x)(\forall y)$ $((x \cup y)' = x' \cap y')$ where $\mathscr{C}$ is the substitution set of x and y. We have assumed that all the sets in the collection acquire their objects from a universal set $\mathbf{U}$.

Inclusion is another relationship between sets for which we shall find much use. Intuitively, a set A is included in a set B if every element of A is also an element of B. This situation we may illustrate with a Venn diagram.

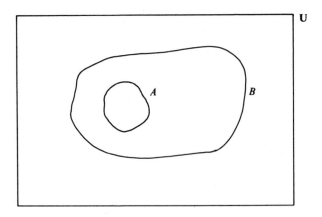

We write $A \subseteq B$ and read "A is included in B", "A is a subset of B" or "B is a superset of A". The fact that $A \subseteq B$ does not exclude the possibility that $A = B$. Thus every set is a subset of itself. Quite often we wish to state that $A \subseteq B$ but $A \neq B$. This means that every object in A is also in B, but B does contain objects which are not in A.

In this event we write $A \subset B$, rather than $(A \subseteq B) \wedge (A \neq B)$, and read "$A$ is a proper subset of B", "A is strictly included in B" or "B is a superset of A". We also write $A \nsubseteq B$ for $\sim(A \subseteq B)$ and $A \not\subset B$ for $\sim(A \subset B)$.

More precisely, $A \subseteq B$ is defined to mean $(\forall x)(x \in A \Rightarrow x \in B)$. If A is $\{x \in \mathbf{U} \,|\, P(x)\}$ and B is $\{x \in \mathbf{U} \,|\, Q(x)\}$ then $A \subseteq B$ means $(\forall x)(P(x) \Rightarrow Q(x))$. This expresses the fact that objects of A which make $P(x)$ true must also make $Q(x)$ true because these objects are in B.

We observe that $(A \subseteq B) \wedge (B \subseteq A)$ means $(\forall x)(x \in A \Rightarrow x \in B) \wedge (\forall x)(x \in B \Rightarrow x \in A)$ which we have already used to mean $A = B$. Thus in a sense inclusion is one half of equality, for $(A \subseteq B) \wedge (B \subseteq A) \Leftrightarrow A = B$. In the first example, we supplied in effect a proof of $A \subseteq B$ and a proof of $B \subseteq A$ in order to deduce $A = B$.

Given a set A, *subset* of A is simply a set X such that $X \subseteq A$. We can consider the set $\{X \,|\, X \subseteq A\}$, whose elements consist of sets which are subsets of A. Let $A = \{0, 1, 2\}$. Then $\{0\}$, $\{1\}$, $\{2\}$, $\{0, 1\}$, $\{1, 2\}$, $\{0, 2\}$ and $\{0, 1, 2\}$ are, all of them, subsets of A. There is one more subset of A, namely the empty set ϕ. In fact, ϕ is a subset of every set, including itself. We may show this quite easily from the definition of the meaning of $\phi \subseteq A$.

Let A be any set. We show that $\phi \subseteq A$. We must prove that $(\forall x)(x \in \phi \Rightarrow x \in A)$. Let $k \in \mathbf{U}$. We must prove $k \in \phi \Rightarrow k \in A$. But this implication is manifestly true because $k \in \phi$ is a false statement – no object is an element of the empty set.

We may observe that if A is a finite set containing just n objects, then the total number of subsets of A is just 2^n. To see this, we consider the method of specifying a subset which consists of indicating, for each element of A, whether or not that element is in the subset. For each element there are just two choices – it is in the subset or it is out. Since there are n elements the number of different subsets obtainable is $\underbrace{2 \times 2 \times 2 \times .. \times 2}_{n \text{ of them}} = 2^n$.

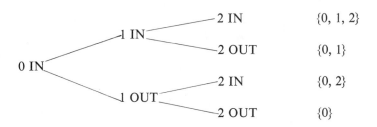

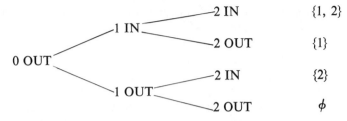

We illustrate the specification of the 2^3 subsets of $\{0, 1, 2\}$.

Exercises 10

1. Let a, b, c, d be the names of four different objects.
   (i) Write down all the subsets of $\{a, b\}$.
   (ii) How many subsets has the set $\{a, b, a, c\}$?
   (iii) Find five subsets A, B, C, D and E of the set $\{a, b, c, d\}$ such that $A \subset B$, $B \subset C$, $C \subset D$ and $D \subseteq E$.
2. Let A and B be subsets of some universal set $\mathbf{U}$. Prove that
   (i) $(A \cap B)' = A' \cup B'$
   (ii) $A \cap A' = \phi$
   (iii) $A \cup A' = \mathbf{U}$
   (iv) $A \setminus A' = A$
   (v) $A \subseteq B \Leftrightarrow A \setminus B = \phi$.
3. Let $A = \{n \in \mathbf{N} \,|\, n \text{ is prime and } n \geqslant 3\}$, let $B = \{n \in \mathbf{N} \,|\, n \text{ is odd}\}$. Prove that $A \subset B$.

Answers 10

1. (i) ϕ, $\{a\}$, $\{b\}$, $\{a, b\}$.
   (ii) The set $\{a, b, a, c,\}$ contains three elements, $2^3 = 8$.
   (iii) $\phi \subset \{a\}$, $\{a\} \subset \{a, b\}$, $\{a, b\} \subset \{a, b, c\}$, $\{a, b, c\} \subset \{a, b, c, d\}$.
2. (i) To prove $(A \cap B)' \subseteq A' \cup B'$, i.e. $(\forall x)(x \in (A \cap B)' \Rightarrow x \in (A' \cup B'))$. Let $k \in \mathbf{U}$. Assume $k \in (A \cap B)'$. Then $k \notin A \cap B$. Thus either $k \notin A$ or $k \notin B$. Thus $k \in A'$ or $k \in B'$ whence $k \in A' \cup B'$. To prove $A' \cup B' \subseteq (A \cap B)'$, i.e. $(\forall x)(x \in A' \cup B') \Rightarrow x \in (A \cap B)')$. Let $k \in \mathbf{U}$. Assume $k \in A' \cup B'$. Then either $k \in A'$ or $k \in B'$. Hence either $k \notin A$ or $k \notin B$. Thus $k \notin A \cap B$ and so $k \in (A \cap B)'$.
   (ii) To prove $A \cap A' \subseteq \phi$, i.e. $(\forall x)(x \in A \cap A' \Rightarrow x \in \phi)$. Let $k \in \mathbf{U}$. Assume $k \in A \cap A'$. Then $k \in A$ and $k \in A'$. Hence from $k \in A'$ we obtain $k \notin A$. We have obtained a contradiction, since $k \in A$ and $\sim(k \in A)$ are both true. Hence the assumption

$k \in A \cap A'$ must be false. The implication $k \in A \cap A' \Rightarrow x \in \phi$ is therefore true.

To prove $\phi \subseteq A \cap A'$. This is true since ϕ is a subset of every set.

(iii) To prove $A \cup A' \subseteq U$; i.e.
$$(\forall x)(x \in A \cup A' \Rightarrow x \in U).$$
Let $k \in U$. Then $k \in A \cup A' \Rightarrow k \in U$ is true since $k \in U$ is true. To prove $U \subseteq A \cup A'$; i.e. $(\forall x)(x \in U \Rightarrow x \in A \cup A')$. Let $k \in U$. Either $k \in A$ or $k \notin A$. Hence either $k \in A$ or $k \in A'$ whence $k \in A \cup A'$.

(iv) To prove $A \setminus A' \subseteq A$. Assume $k \in A \setminus A'$. Then $k \in A$ (and $k \notin A'$). To prove $A \subseteq A \setminus A'$. Assume $k \in A$. Then $k \notin A'$ so $k \in A \setminus A'$.

(v) We must prove $(A \subseteq B \Rightarrow A \setminus B = \phi) \wedge (A \setminus B = \phi \Rightarrow A \subseteq B)$. First we prove $A \subseteq B \Rightarrow A \setminus B = \phi$. Assume $A \subseteq B$. We must prove $A \setminus B = \phi$. Of course $\phi \subseteq A \setminus B$, we must show $A \setminus B \subseteq \phi$. Assume $k \in A \setminus B$. Then $k \in A$ and $k \notin B$. Thus $k \in A$ is true and $k \in B$ is false. Hence $k \in A \Rightarrow k \in B$ is false. But $A \subseteq B$, i.e. $(\forall x)(x \in A \Rightarrow x \in B)$ and in particular $k \in A \Rightarrow k \in B$. Hence we have a contradiction. We have made two assumptions, at least one of which must be false. If $A \subseteq B$ is false then $A \subseteq B \Rightarrow A \setminus B = \phi$ is true by truth tables. If $k \in A \setminus B$ is false then $k \in A \setminus B \Rightarrow k \in \phi$ is true so $A \setminus B \subseteq \phi$ is true, whence $A \subseteq B \Rightarrow A \setminus B = \phi$ is true. Thus we have proved $A \subseteq B \Rightarrow A \setminus B = \phi$. Next we prove $(A \setminus B = \phi) \Rightarrow A \subseteq B$. Assume $A \setminus B = \phi$. We must prove $A \subseteq B$. Assume $k \in A$. Were $k \notin B$ then would $k \in A \setminus B$ which cannot be true since $A \setminus B = \phi$, hence $k \in B$. This proves $A \subseteq B$ and so we have proved $(A \setminus B = \phi) \Rightarrow A \subseteq B$. Hence, combining the two proofs, $A \subseteq B \Leftrightarrow A \setminus B = \phi$.

We may note that the above proofs rely very little on intuition about the entities involved. They rely on the understanding of the language involved and follow rather mechanically the methods we have described for proving statements. The advantage of such mechanical methods is that they suggest the way ahead when our intuition fails.

3. To prove $A \subseteq B$. Let $k \in U$ and assume $k \in A$. Then k is prime and $k \geqslant 3$. A prime number is divisible only by itself and 1, so k is not divisible by 2 and hence is odd. Thus $k \in B$. We now have to show that A is a proper subset of B, i.e. that there is an element of B which is not in A. Consider the number 15. Certainly 15 is odd so

$15 \in B$. But 3 divides 15 and 3 is neither 1 nor 15 so 15 is not prime; thus $15 \notin A$. We conclude that $A \subset B$.

§5 The Power Set

Given a set A we can define a new set, called the *power set of A* and denoted $P(A)$, by

$$P(A) = \{x \mid x \subseteq A\}.$$

Thus the power set of A is the set of all subsets of A.

For example the power set of $\{0, 1\}$ is the set $\{\phi, \{0\}, \{1\}, \{0, 1\}\}$. We have listed the subsets of $\{0, 1\}$ within curly brackets in order to describe $P(\{0, 1\})$.

If A is a finite set containing n elements then $P(A)$ is a finite set containing 2^n elements. This is merely a restatement of the observation made in §4.

CHAPTER 6

RELATIONSHIPS

§1 Relationships as Objects

The basic idea of a relationship is that we have a collection of objects, some of which are related to others in some way. Thus 3 and 5 are related by one being greater than the other; $\frac{1}{2}$ and $\frac{2}{4}$ are related by one being equal to the other; people may be related by one being married to the other, one being of the same nationality as the other, one being the son of the other. These relationships have names, but there are many relationships between objects which are not sufficiently important to have names.

We should like to be able to deal with all possible relationships; it is also right that a thing which can be given a name should be an object and we shall define what a relationship is in such a way that it is an object.

If one wishes to explain a relationship to another person and there is no common name for the relationship, then one may accomplish this by simply specifying which objects are related to which by the relationship. Often this is not possible because of the number of objects involved. In this event, one must give the person a rule which he can use to decide for any two objects a and b, whether a stand in this relationship to b or not. This rule takes the form of a binary predicate $P(x, y)$ such that a stands in this relationship to b if $P(a, b)$ is true and a does not stand in this relationship to b if $P(a, b)$ is false. Of course we will have in mind substitution sets for x and y for which $P(x, y)$ makes sense. It is only reasonable that we should have the objects in hand before we decide which is to be related to which.

DEFINITION. Let A and B be sets. A relation on $A \times B$ is a subset of $A \times B$.

Thus a *relation* is an object. In fact, it is a set whose elements are of the form (a, b). Let R be a relation on $A \times B$. This means that R is a set of ordered pairs whose first part is an element of A and whose second is an element of B. Thus R specifies a relationship between objects in A and objects in B. Let $(a, b) \in A \times B$. If $(a, b) \in R$ then a is related to b by R, while if $(a, b) \notin R$ then a is not related to b by R.

Since R is a set it can be defined by a predicate. Since R is a set of ordered pairs, the predicate will be a binary predicate. Let $P(x, y)$ be a binary predicate and let A and B be the substitution sets of x and y respectively. The relation described by $P(x, y)$ is the set $\{(x, y) \in A \times B \mid P(x, y)\}$.

EXAMPLES

Let F be a set of women and let M be a set of men. We may define the relations

$$\{(x, y) \in F \times M \mid x \text{ is the mother of } y\}$$
$$\{(x, y) \in F \times M \mid x \text{ is married to } y\}$$
$$\{(x, y) \in F \times M \mid x \text{ is the sister of } y\}$$
$$\{(x, y) \in F \times M \mid x \text{ and } y \text{ have the same father}\})$$
$$\{(x, y) \in F \times M \mid y \text{ went to school with } x\}.$$

Frequently we wish to discuss relationships between objects in the same set A. The relation in such a case will be a subset of $A \times A$. Thus for example:

$$\{(x, y) \in M \times M \mid x \text{ is the brother of } y\}$$
$$\{(x, y) \in \mathbf{N} \times \mathbf{N} \mid x \text{ exactly divides } y\}$$
$$\{(x, y) \in \mathbf{R} \times \mathbf{R} \mid x > y\}$$
$$\{(x, y) \in \mathbf{R} \times \mathbf{R} \mid x + y = 1\}.$$

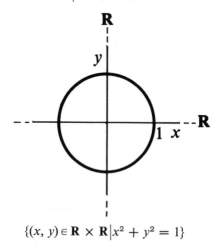

$$\{(x, y) \in \mathbf{R} \times \mathbf{R} \mid x^2 + y^2 = 1\}$$

If we confine ourselves to relations on $\mathbf{R} \times \mathbf{R}$, then we can draw pictures of relations using the cartesian plane. Let $R \subseteq \mathbf{R} \times \mathbf{R}$. Then we draw our picture by inking in the points of the plane which have co-ordinates (a, b) such that $(a, b) \in R$. Of course we have also inked in the axes for reference, and it is also clear that having only a finite amount of paper we may only be able to draw part of the relation.

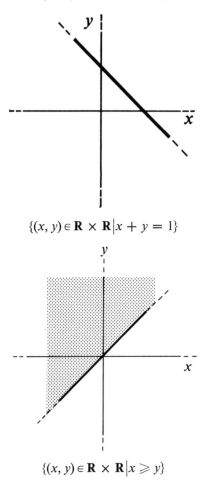

$$\{(x, y) \in \mathbf{R} \times \mathbf{R} \mid x + y = 1\}$$

$$\{(x, y) \in \mathbf{R} \times \mathbf{R} \mid x \geqslant y\}$$

The picture may consist of lines, whole areas or isolated points. Such pictures are useful for gaining an intuitive appreciation of a relation, but this medium is best for representing relations which yield lines rather than areas or isolated points.

Given a relation R on $A \times B$ we can define another relation called the *inverse relation* of R.

DEFINITION. Let R be a relation on $A \times B$. We define a relation on $B \times A$ called the *inverse relation* of R, which we denote R^{-1}, by

$$R^{-1} = \{(y, x) \in B \times A \,|\, (x, y) \in R\}.$$

Thus if $R = \{(x, y) \in F \times M \,|\, x$ is the mother of $y\}$, then $R^{-1} = \{(y, x) \in M \times F \,|\, x$ is the mother of $y\}$. Of course there is an alternative method of expressing R^{-1}, namely $\{(y, x) \in M \times F \,|\, y$ is the son of $x\}$.

By and large we have grown accustomed to using x to mark the first space in an ordered pair and y to mark the second. These variables simply distinguish two kinds of spaces, one sort of space to be filled by an object from one set and the other sort to be filled by an object from another. There is no reason why we should not interchange the labelling of these two kinds of spaces.

Thus $R^{-1} = \{(y, x) \in M \times F \,|\, y$ is the son of $x\}$ can also be written $\{(x, y) \in M \times F \,|\, x$ is the son of $y\}$. We observe that the substitution sets are attached to the spaces, not the variables which mark the spaces. Hence when a variable is used as a space marker, it acquires the substitution set appropriate to the space which it marks. Thus, in the example, we took the spaces marked y and changed the marker y to x. The substitution set of y was M and so x acquires M as its substitution set. At the same time the spaces which were marked x and had substitution set F have now been marked y and so the substitution set of y is now F. We must interchange the markers simultaneously in order that the two different kinds of spaces will be marked with different variables.

We observe that "x is the mother of y" and "y is the son of x" where $x \in F$ and $y \in M$ are the same predicate in x and y in the sense that the first is true when a replaces x and b replaces y if and only if the second is true when a replaces x and b replaces y. However, x occurs before y in the first predicate and this order is reversed in the second. Usually we prefer to write

$$R = \{(x, y) \in F \times M \,|\, x \text{ is the mother of } y\}$$

and

$$R^{-1} = \{(y, x) \in M \times F \,|\, y \text{ is the son of } x\}$$

thus preserving in the predicate the order of variables as they occur in the ordered pair. Thus where possible we like to say how the first part of the ordered pair is related to the second.

Let us combine this preference for keeping the variables in the same order in both R and R^{-1} with the preference for keeping x before y. By interchanging variables we obtain

$$R^{-1} = \{(x, y) \in M \times F \mid x \text{ is the son of } y\}.$$

Now, given $R = \{(x, y) \in F \times M \mid x \text{ is the mother of } y\}$ we may apply the definition of inverse relation to obtain $R^{-1} = \{(y, x) \in M \times F \mid x \text{ is the mother of } y\}$. This is perfectly good as it stands, but if we indulge our preferences we could render R^{-1} as $\{(x, y) \in M \times F \mid x \text{ is the son of } y\}$. We see that this indulgence necessitates our finding a suitable alternative predicate instead of "x is the mother of y". This is not always possible in general, but where it is possible it is often done.

EXAMPLE. Let $R = \{(x, y) \in F \times M \mid x \text{ is the sister of } y\}$. From the definition, $R^{-1} = \{(y, x) \in M \times F \mid x \text{ is the sister of } y\}$. Indulging our preferences,

$$R^{-1} = \{(x, y) \in M \times F \mid x \text{ is the brother of } y\}.$$

Let $R = \{(x, y) \in \mathbf{R} \times \mathbf{R} \mid x \leqslant y\}$. From the definition, $R^{-1} = \{(y, x) \in \mathbf{R} \times \mathbf{R} \mid x \leqslant y\}$. Indulging our preferences, $R^{-1} = \{(x, y) \in \mathbf{R} \times \mathbf{R} \mid x \geqslant y\}$.

The preferred form of R^{-1} also depends on the nature of R. For example, it is quite correct to define $R = \{(x, y) \in \mathbf{R} \times \mathbf{R} \mid 2x + y = 1\}$ and to define $R^{-1} = \{(y, x) \in \mathbf{R} \times \mathbf{R} \mid 2x + y = 1\}$. There are good reasons why we would in fact prefer R to be given as $\{(x, y) \in \mathbf{R} \times \mathbf{R} \mid y = 1 - 2x\}$ and R^{-1} as $\{(x, y) \in \mathbf{R} \times \mathbf{R} \mid y = \frac{1}{2}(1 - x)\}$. We obtain this form of R^{-1} by arranging the predicate $y = 1 - 2x$ in the equivalent form $x = \frac{1}{2}(1 - y)$ and then interchanging variables.

We have mentioned that there are predicates which are not sufficiently indulgent for such purposes. An example is

$$R = \{(x, y) \in \mathbf{R} \times \mathbf{R} \mid 2x^2 + y^2 = 1\}.$$

The predicate $2x^2 + y^2 = 1$ cannot be expressed in the form $y = \ldots$ (an expression in x). The inverse of this relation is by definition $R^{-1} = \{(y, x) \in \mathbf{R} \times \mathbf{R} \mid 2x^2 + y^2 = 1\}$. We can always interchange variables and obtain R^{-1} in the form $\{(x, y) \in \mathbf{R} \times \mathbf{R} \mid 2y^2 + x^2 = 1\}$ even though we still cannot write the predicate in the form $y = $ (an expression in x).

We can draw instructive pictures of some relations and their inverses. Thus if $R = \{(x, y) \in \mathbf{R} \times \mathbf{R} \mid y = x^2\}$ we have the following picture.

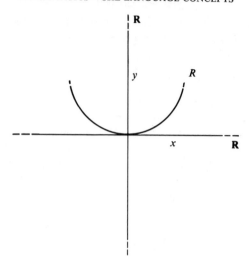

The picture of $R^{-1} = \{(y, x) \in \mathbf{R} \times \mathbf{R} \mid y = x^2\}$ may be drawn thus.

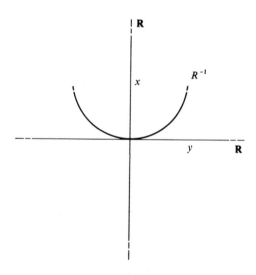

Thus we obtain the same picture, since we have exchanged axes so that the first variable still takes objects from the horizontal line. It is better if we exchange variables, obtaining $R^{-1} = \{(x, y) \in \mathbf{R} \times \mathbf{R} \mid x$

$= y^2\}$. Then, with the first variable still taking objects from the horizontal line, we obtain:

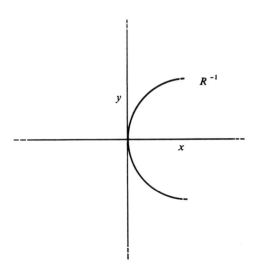

If we draw both $R = \{(x, y) \in \mathbf{R} \times \mathbf{R} \mid y = x^2\}$ and $R^{-1} = \{(x, y) \in \mathbf{R} \times \mathbf{R} \mid x = y^2\}$ on the same diagram we obtain:

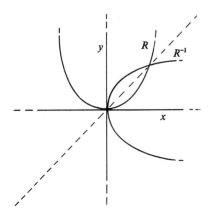

We can now see that R^{-1} is the mirror image of R in the line $y = x$. This is probably the most instructive picture of the situation. In general, if we have a picture of R we can obtain a picture of R^{-1} simply by taking the mirror image of R in the line $y = x$.

Since a relation on $A \times B$ is just a subset of $A \times B$, it follows that the set of all relations on $A \times B$ is the set of all subsets of $A \times B$, i.e. the power set of $A \times B$. In particular ϕ and $A \times B$ are subsets of $A \times B$ and hence are relations. The relation ϕ is a relation such that no object of A is related to any object of B. The relation $A \times B$ is such that every object in A is related to every object in B.

Exercises 11

1. For the following relations the substitution set of x and y is a set M of men. State the inverse relations (a) by direct reference to the definition and (b) in the preferred form $\{(x, y) \in M \times M \,|\, P(x, y)\}$ where x comes before y in the predicate $P(x, y)$.

$R =$
   (i) $\{(x, y) \in M \times M \,|\, x$ is older than $y\}$
   (ii) $\{(x, y) \in M \times M \,|\, x$ has shaken hands with $y\}$
   (iii) $\{(x, y) \in M \times M \,|\, x$ is at least as tall as $y\}$
   (iv) $\{(x, y) \in M \times M \,|\, x$ knows the name of $y\}$
   (v) $\{(x, y) \in M \times M \,|\, x$ is the brother of $y\}$.

2. For the following relations, state the inverse relation (a) by direct application of the definition, (b) in the preferred form obtained by interchange of variables. Draw the relation and its inverse in the same picture.

$R =$
   (i) $\{(x, y) \in \mathbf{R} \times \mathbf{R} \,|\, y = x^3\}$
   (ii) $\{(x, y) \in \mathbf{R} \times \mathbf{R} \,|\, y = 2x - 1\}$
   (iii) $\left\{(x, y) \in \mathbf{R} \times \mathbf{R} \,\middle|\, y - 1 = \dfrac{1}{x - 2}\right\}$
   (iv) $\left\{(x, y) \in \mathbf{R} \times \mathbf{R} \,\middle|\, \dfrac{x^2}{9} + \dfrac{y^2}{16} = 25\right\}$
   (v) $\{(x, y) \in \mathbf{R} \times \mathbf{R} \,|\, x^2 + y^2 = 1\}$.

Answers 11

1. (i) (a) $\{(y, x) \in M \times M \,|\, x$ is older than $y\}$.
      (b) $\{(x, y) \in M \times M \,|\, x$ is younger than $y\}$.

   (ii) (a) $\{(y, x) \in M \times M \,|\, x$ has shaken hands with $y\}$.
       (b) $\{(x, y) \in M \times M \,|\, x$ has shaken hands with $y\}$.

   (iii) (a) $\{(y, x) \in M \times M \,|\, x$ is at least as tall as $y\}$.
        (b) $\{(x, y) \in M \times M \,|\, x$ is at most as tall as $y\}$.

(iv) (a) $\{(y, x) \in M \times M \mid x$ knows the name of $y\}$.

   (b) $\{(x, y) \in M \times M \mid$ The name of x is known by $y\}$.

(v) (a) $\{(y, x) \in M \times M \mid x$ is the brother of $y\}$.

   (b) $\{(x, y) \in M \times M \mid x$ is the brother of $y\}$.

2. (i) (a) $\{(y, x) \in \mathbf{R} \times \mathbf{R} \mid y = x^3\}$

     (b) $\{(x, y) \in \mathbf{R} \times \mathbf{R} \mid x = y^3\}$

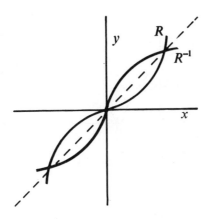

(ii) (a) $\{(y, x) \times \mid y = 2x - 1\}$

    (b) $\{(x, y) \times \mid x = 2y - 1\}$

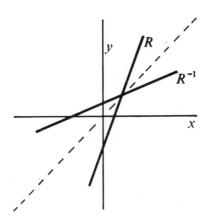

(iii) (a) $\left\{(y, x) \times \left| y- = \dfrac{1}{x-2}\right.\right\}$ (b) $\left\{(x, y) \times \left| x-1 = \dfrac{1}{y-2}\right.\right\}$

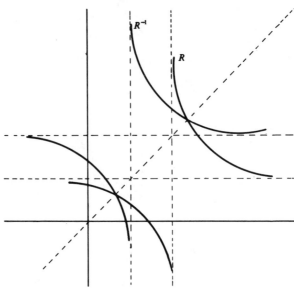

(iv) (a) $\left\{(y, x) \in \mathbf{R} \times \mathbf{R} \left| \dfrac{x^2}{9} + \dfrac{y^2}{16} = 25\right.\right\}$

(b) $\left\{(x, y) \in \mathbf{R} \times \mathbf{R} \left| \dfrac{x^2}{16} + \dfrac{y^2}{9} = 25\right.\right\}$

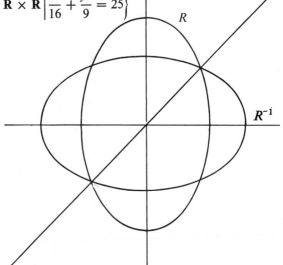

(v) (a) $\{(y, x) \in \mathbf{R} \times \mathbf{R} \mid x^2 + y^2 = 1\}$
(b) $\{(x, y) \in \mathbf{R} \times \mathbf{R} \mid x^2 + y^2 = 1\}$.

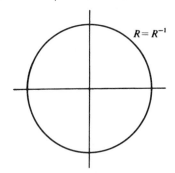

$R = R^{-1}$

§2 Properties which Characterise Relations

Let A be a set. We consider the relations which can be defined on $A \times A$. The set of all relations on $A \times A$ is just the set of all subsets of $A \times A$ and is thus simply the power set of $A \times A$, i.e. $P(A \times A)$.

Consider the predicate $(x, x) \in R$, where A is the substitution set of x and $P(A \times A)$ is the substitution set of R. By quantifying the variable x we obtain $(\forall x)((x, x) \in R)$ which is now a predicate in the single variable R. If we substitute a relation from $P(A \times A)$ for the variable R in this predicate then the resulting statement may be true or false. Thus the predicate effectively divides the elements of $P(A \times A)$ into two kinds—those for which it is true and those for which it is false. We have a name for those relations which make $(\forall x)((x, x) \in R)$ true; they are described as *reflexive*.

Let k be a reflexive relation on $A \times A$ so that $(\forall x)((x, x) \in k)$ is true. Then the statement $(\forall x)((x, x) \in k)$ tells us that the relation k is such that every element of A stands in the relationship k to itself.

Let the elements of A be people. Then the following are examples of reflexive relations on $A \times A$.

$$\{(x, y) \in A \times A \mid x \text{ is of the same nationality as } y\}$$
$$\{(x, y) \in A \times A \mid x \text{ is at least as tall as } y\}$$
$$\{(x, y) \in A \times A \mid x \text{ knows the name of } y\}.$$

The following relations on $\mathbf{R} \times \mathbf{R}$ are reflexive.

$$\{(x, y) \times \mathbf{R} \times \mathbf{R} \mid x \leqslant y\}$$
$$\{(x, y) \in \mathbf{R} \times \mathbf{R} \mid x^2 = y^2\}$$
$$\{(x, y) \in \mathbf{R} \times \mathbf{R} \mid x = y\}.$$

A reflexive relation k on $\mathbf{R} \times \mathbf{R}$ can be recognised by its picture. Since for every real number a we must have $(a, a) \in k$, it follows that the picture must always *include* the line $y = x$.

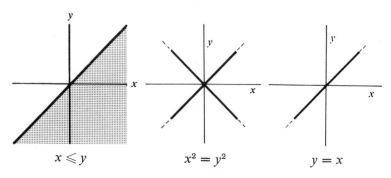

$$x \leqslant y \qquad\qquad x^2 = y^2 \qquad\qquad y = x$$

We observe that $(\forall x) ((x, x) \in k)$ tells us that the relation k contains certain ordered pairs. The larger the set A, the more ordered pairs k must contain in order to be reflexive. Consider the set E defined by

$$\{(x, y) \in \mathbf{R}_0^+ \times \mathbf{R}_0^+ \,|\, x = y\},$$

where $\mathbf{R}_0^+$ is the set of non-negative reals. Of course E is a reflexive relation on $\mathbf{R}_0^+ \times \mathbf{R}_0^+$, but it is evidently also a subset of $\mathbf{R} \times \mathbf{R}$ and hence is a relation on $\mathbf{R} \times \mathbf{R}$. However, it is not a reflexive relation on $\mathbf{R} \times \mathbf{R}$ since this would necessitate ordered pairs like $(-3, -3)$ being elements of E which they are not.

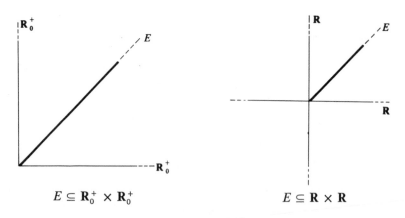

$$E \subseteq \mathbf{R}_0^+ \times \mathbf{R}_0^+ \qquad\qquad E \subseteq \mathbf{R} \times \mathbf{R}$$

DEFINITION. A relation k on a set $A \times A$ is *reflexive* if and only if $(\forall x)((x, x) \in k)$ is true, where A is the substitution set of x.

Consider next the predicate $(\forall x)(\forall y)((x, y) \in R \Rightarrow (y, x) \in R)$ where A is the substitution set of both x and y and $P(A \times A)$ the substitution set of R. A relation on $A \times A$ which makes this predicate into a true statement is described as *symmetric*. Let k be a symmetric relation on $A \times A$ so that $(\forall x)(\forall y)((x, y) \in k \Leftrightarrow (y, x) \in k)$ is true. This statement does not mean that k contains any particular ordered pairs, but if it should contain the pair (a, b) then it must also contain the pair (b, a).

Let the elements of A be people, then the following are examples of symmetric relations on $A \times A$.

$$\{(x, y) \in A \times A \,|\, x \text{ is married to } y\}$$
$$\{(x, y) \in A \times A \,|\, x \text{ is the cousin of } y\}$$
$$\{(x, y) \in A \times A \,|\, x \text{ has the same parents as } y\}.$$

If k is a symmetric relation on $\mathbf{R} \times \mathbf{R}$ and if $(a, b) \in k$, then $(b, a) \in k$. But (b, a) is the mirror image of (a, b) in the line $y = x$. Hence the picture of a symmetric relation on $\mathbf{R} \times \mathbf{R}$ must be symmetric about the line $y = x$.

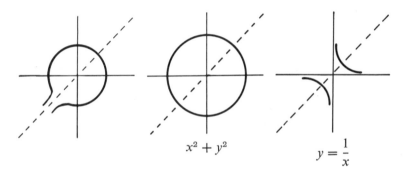

$$x^2 + y^2 \qquad\qquad y = \frac{1}{x}$$

DEFINITION. A relation k on a set $A \times A$ is *symmetric* if and only if $(\forall x)(\forall y)((x, y) \in k \Rightarrow (y, x) \in k)$ is true, where A is the substitution set of both x and y.

Consider the predicate

$$(\forall x)(\forall y)(\forall z)((x, y) \in R \wedge (y, z) \in R \Rightarrow (x, z) \in R)$$

where A is the substitution set of x, y and z and $P(A \times A)$ is the substitution set of R. A relation on $A \times A$ which makes this predicate into a true statement is described as *transitive*.

Let k be a transitive relation on $A \times A$. Then if an element $a \in A$ is related by k to an element b, and if also b is related by k to a third element c then necessarily the first element a is related by k to the third element c.

Let the elements of A be people; then the following relations on $A \times A$ are transitive.

$$\{(x, y) \in A \times A \mid x \text{ is older than } y\}$$
$$\{(x, y) \in A \times A \mid x \text{ is a blood relation of } y\}.$$

Also

$$\{(x, y) \in \mathbf{R} \times \mathbf{R} \mid x \geqslant y\},$$
$$\{(x, y) \in \mathbf{R} \times \mathbf{R} \mid x = y\} \text{ and, for any set } A,$$
$$\{(x, y) \in P(A) \times P(A) \mid x \subseteq y\} \text{ are transitive.}$$

We may illustrate this last relation using a Venn diagram.

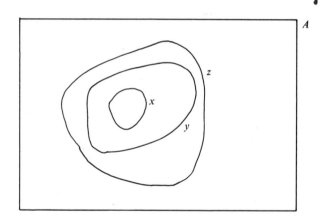

The universal set is A; x, y and z are subsets of A, and thus elements of $P(A)$. Given that $x \subseteq y$ and $y \subseteq z$ it is evident from the picture that $x \subseteq z$.

Consider the predicate

$$(\forall x)(\forall y)((x, y) \in R \Rightarrow (y, x) \notin R)$$

where A is the substitution set of both x and y and $P(A \times A)$ is the substitution set of R. A relation on $A \times A$ which makes this predicate into a true statement is described as *antisymmetric*. Let k be an anti-symmetric relation on $A \times A$. Then if (a, b) is an element of k, (b, a) is not an element of k.

If A is a set of people then the relations

$$\{(x, y) \in A \times A \,|\, x \text{ is the wife of } y\}$$
$$\{(x, y) \in A \times A \,|\, x \text{ is the father of } y\}$$
$$\{(x, y) \in A \times A \,|\, x \text{ is the son of } y\}$$

are antisymmetric.

Also

$$\{(x, y) \in \mathbf{R} \times \mathbf{R} \,|\, < y\} \text{ and, for any set A,}$$
$$\{(x, y) \in P(A) \times P(A) \,|\, x \subset y\} \text{ are antisymmetric.}$$

If k is an antisymmetric relation on $\mathbf{R} \times \mathbf{R}$ and $(a, b) \in k$, then (b, a), the mirror image of (a, b), cannot be an element of k. In particular, no ordered pair (a, a) can be an element of k since assuming $(a, a) \in k$, by antisymmetry

$$(a, a) \in k \Rightarrow (a, a) \notin k,$$

and we deduce $(a, a) \notin k$ contradicting the assumption that $(a, a) \in k$. Thus we have in fact $(\forall x)((x, x) \notin k)$.

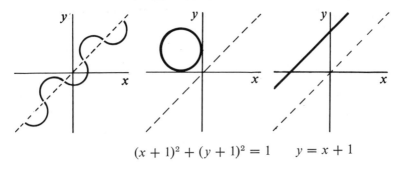

$$(x + 1)^2 + (y + 1)^2 = 1 \qquad y = x + 1$$

Exercises 12

Let A be a set and let R and S be relations on $A \times A$. The complements of R and S are taken with respect to $A \times A$.

1. Show that the relation $\phi \subseteq A \times A$ is symmetric, transitive and antisymmetric. Show that if ϕ is reflexive then $A = \phi$.
2. Show that if R is both symmetric and antisymmetric then $R = \phi$.
3. Show that if R is reflexive and $R \subseteq S$ then S is reflexive. Also that if R is antisymmetric and $S \subseteq R$ then S is antisymmetric.
4. Show that if R is antisymmetric then R' is reflexive.
5. Show that if R is transitive and symmetric and every element of A is related by R to at least one object in A then R is reflexive.

6. Show that R is symmetric if and only if $R = R^{-1}$.
7. Show that R is antisymmetric if and only if $R \cap R^{-1} = \phi$.
8. Show that for any relation R, the relation $R \cup R^{-1}$ is symmetric and the relation $R \setminus R^{-1}$ is antisymmetric.
9. Show that if R and S are reflexive then so are the relations $R \cup S$ and $R \cap S$, but $R \setminus S$ and R' cannot be reflexive.
10. Show that if R is antisymmetric then
    (a) $R \cap S$ and $R \setminus S$ must be antisymmetric,
    (b) R' cannot be antisymmetric, unless $A = \phi$,
    (c) $R \cup S$ need not *necessarily* be antisymmetric, even if S is also antisymmetric.
11. Show that if R and S are symmetric, then so are $R \cup S$, $R \cap S$, $R \setminus S$ and R'.
12. Show that if R is transitive, then so is R^{-1}.
13. Prove that the relation $\{(x, y) \in P(A) \times P(A) | x \subseteq y\}$ is transitive, using the definition of subset.

Answers 12

1. In order to prove ϕ symmetric, we must prove

$$(\forall x)(\forall y)((x, y) \in \phi \Rightarrow (y, x) \in \phi).$$

Let a and b be arbitrary elements of A. We have to prove that $(a, b) \in \phi \Rightarrow (b, a) \in \phi$. But this implication is true since $(a, b) \in \phi$ is false. The proofs that ϕ is transitive and antisymmetric are very similar.

Assume ϕ is reflexive, then $(\forall x)((x, x) \in \phi)$. Suppose $A \neq \phi$, then let a be an element of A. Therefore $(a, a) \in \phi$, which is false. Hence the supposition was false and $A = \phi$.

2. By contradiction. Assume that R is both symmetric and antisymmetric but that $R \neq \phi$. Since $R \neq \phi$, let $(a, b) \in R$. By symmetry $(\forall x)(\forall y)((x, y) \in R \Rightarrow (y, x) \in R)$ so in particular $(a, b) \in R \Rightarrow (b, a) \in R$. Hence $(b, a) \in R$. By antisymmetry $(\forall x)(\forall y)((x, y) \in R \Rightarrow (y, x) \notin R)$ so in particular $(a, b) \in R \Rightarrow (b, a) \notin R$. Hence $(b, a) \notin R$, contradiction.

3. Define Δ to be the set $\{(a, a) | a \in A\}$. Then R is reflexive, hence $\Delta \subseteq R$. But $R \subseteq S$. Hence $\Delta \subseteq S$ (using the transitivity of $\subseteq$). Hence S is reflexive.

To prove $(\forall x)(\forall y)((x, y) \in S \Rightarrow (y, x) \notin S)$ assuming $(\forall x)(\forall y)((x, y) \in R \Rightarrow (y, x) \notin R)$ and $S \subseteq R$. Let $a, b \in A$. We prove $(a, b) \in S \Rightarrow (b, a) \notin S$ directly. Assume $(a, b) \in S$. Now $S \subseteq R$ so $(a, b) \in R$. Now $(\forall x)(\forall y)((x, y) \in R \Rightarrow (y, x) \notin R)$ yields in particular

$(a, b) \in R \Rightarrow (b, a) \notin R$. Hence $(b, a) \notin R$. Since S is contained in R, $(b, a) \notin S$.

4. To prove $(\forall x)((x, x) \in R')$ under the assumption $(\forall x)(\forall y)((x, y) \in R \Rightarrow (y, x) \notin R)$. Let $a \in A$. We prove $(a, a) \in R'$ by contradiction. Suppose $(a, a) \notin R'$. Then $(a, a) \in R$. Since R is antisymmetric we have in particular $(a, a) \in R \Rightarrow (a, a) \notin R$. Hence $(a, a) \notin R$, contradiction. Thus the supposition $(a, a) \notin R'$ was false; so $(a, a) \in R'$.

5. To prove $(\forall x)((x, x) \in R)$ under the assumption that R is transitive, symmetric, and that every element of A is related by R to at least one element of A. Let $a \in A$. To prove $(a, a) \in R$. Now let b be one of the elements of A to which a is related; our assumption guarantees the existence of such a b. Then $(a, b) \in R$. By symmetry $(b, a) \in R$. By transitivity $(a, b) \in R \wedge (b, a) \in R$ yields $(a, a) \in R$.

6. Firstly we prove $R = R^{-1}$ under the assumption that R is symmetric.

    (a) To show $R \subseteq R^{-1}$. Let $(a, b) \in R$. By the symmetry of R, $(b, a) \in R$. By definition of R^{-1} since $(b, a) \in R$ we must have $(a, b) \in R^{-1}$.

    (b) To show $R^{-1} \subseteq R$. Let $(a, b) \in R^{-1}$. Then using the definition of R^{-1}, $(b, a) \in R$. By the symmetry of R, $(a, b) \in R$. (NOTE, the inverse of R^{-1} is R).

    Secondly we prove that R is symmetric under the assumption $R = R^{-1}$. We have to prove $(\forall x)(\forall y)((x, y) \in R \Rightarrow (y, x) \in R)$. Let $a, b \in A$ and assume $(a, b) \in R$. By definition of R^{-1}, $(b, a) \in R^{-1}$. Since $R^{-1} = R$, $(b, a) \in R$.

7. Firstly we prove $R \cap R^{-1} = \phi$ on the assumption that R is antisymmetric. By contradiction. Suppose $R \cap R^{-1} \neq \phi$. Then let $(a, b) \in R \cap R^{-1}$. Then $(a, b) \in R$ and $(a, b) \notin R^{-1}$. From $(a, b) \in R$ and the antisymmetry of R it follows that $(b, a) \notin R$. From $(a, b) \in R^{-1}$ it follows from the definition of R^{-1} that $(b, a) \in R$, contradiction. Next, let $R \cap R^{-1} = \phi$. We prove $(\forall x)(\forall y)(x, y) \in R \Rightarrow (y, x) \notin R)$. Let $(a, b) \in R$. Then $(b, a) \in R^{-1}$. Were $(b, a) \in R$ then we would have $(b, a) \in R \cap R^{-1}$ which is not possible because $R \cap R^{-1} = \phi$. Hence $(b, a) \notin R$ as required.

8. Assume $(a, b) \in R \cup R^{-1}$. Hence either $(a, b) \in R$ whence $(b, a) \in R^{-1}$ so that $(b, a) \in R \cup R^{-1}$, or else $(a, b) \in R^{-1}$ whence $(b, a) \in R$ so that again $(b, a) \in R \cup R^{-1}$. Hence

$$(\forall x)(\forall y)((x, y) \in R \cup R^{-1} \Rightarrow (y, x) \in R \cup R^{-1}).$$

Next we have to prove $(\forall x)(\forall y)((x, y) \in R \setminus R^{-1} \Rightarrow (y, x) \notin R \setminus R^{-1})$. Let $(a, b) \in R \setminus R^{-1}$. Then $(a, b) \in R$ and $(a, b) \notin R^{-1}$. Were $(b, a) \in R$ then would $(a, b) \in R^{-1}$, so $(b, a) \notin R$. Thus $(b, a) \notin R \setminus R^{-1}$.

9. Define $\Delta = \{(a, a) | a \in A\}$. A relation on $A \times A$ is reflexive if and only if it is a superset of Δ. Hence if R and S are reflexive then $\Delta \subseteq R$ and $\Delta \subseteq S$ whence $\Delta \subseteq R \cup S$ and $\Delta \subseteq R \cap S$ but $\Delta \cap (R \setminus S) = \Delta \cap (S \setminus R) = \Delta \cap R' = \phi$.

10. (a) Assume $(a, b) \in R \cap S$. Then $(a, b) \in R$, and R is antisymmetric so $(b, a) \notin R$. Therefore $(b, a) \notin R \cap S$. This proves

$$(\forall x)(\forall y)((x, y) \in R \cap S \Rightarrow (y, x) \notin R \cap S).$$

Next, assume $(a, b) \in R \setminus S$. Again $(b, a) \notin R$ and so $(b, a) \notin R \setminus S$.

(b) Since R is antisymmetric, $\Delta \subseteq R'$. See 9.

(c) If $R = \{(-1, 0)\}$ and $S = \{(0, -1)\}$ then R and S are antisymmetric but $R \cup S$ is symmetric.

11. Assume $(a, b) \in R \cup S$. Then *either* $(a, b) \in R$ whence, by the symmetry of R, $(b, a) \in R$ and so $(b, a) \in R \cup S$ *or* $(a, b) \in S$ whence, by the symmetry of S, $(b, a) \in S$ and so $(b, a) \in R \cup S$. In either event we have $(b, a) \in R \cup S$.

Assume $(a, b) \in R \cap S$, then $(a, b) \in R$ and $(a, b) \in S$ whence by the symmetry of R and S, $(b, a) \in R$ and $(b, a) \in S$ and so $(b, a) \in R \cap S$.

Assume $(a, b) \in R \setminus S$, then $(a, b) \in R$ and $(a, b) \notin S$. By the symmetry of R, $(b, a) \in R$. Were $(b, a) \in S$ then would $(a, b) \in S$ by the symmetry of S, which is not so. Hence $(b, a) \notin S$. Thence $(b, a) \in R \setminus S$.

Assume $(a, b) \in R'$, then $(a, b) \notin R$. Were $(b, a) \in R$ then by the symmetry of R would $(a, b) \in R$ which is not so. Hence $(b, a) \notin R$ and so $(b, a) \in R'$.

12. Given that R is transitive we have to prove

$$(\forall x)(\forall y)(\forall z)((x, y) \in R^{-1} \wedge (y, z) \in R^{-1} \Rightarrow (x, z) \in R^{-1}).$$

Let a, b and c be elements of A and assume $(a, b) \in R^{-1}$ and $(b, c) \in R^{-1}$. By the definition of R^{-1} we have $(b, a) \in R$ and $(c, b) \in R$. But since R is transitive $(c, b) \in R \wedge (b, a) \in R \Rightarrow (c, a) \in R$. Hence $(c, a) \in R$. By definition of R^{-1}, $(a, c) \in R^{-1}$. This proves the implication $(a, b) \in R^{-1} \wedge (b, c) \in R^{-1} \Rightarrow (a, c) \in R^{-1}$ as required.

13. We have to prove

$$(\forall x)(\forall y)(\forall z)((x, y) \in k \wedge (y, z) \in k \Rightarrow (x, y) \in k)$$

where $k = \{(x, y) \in P(A) \times P(A) | x \subseteq y\}$
$= \{(x, y) \in P(A) \times P(A) | (\forall w)(w \in x \Rightarrow w \in y)\}$.

The substitution set of x, y, z is $P(A)$ and that of w is A. Let a, b, c $\in P(A)$. To prove $(a, b) \in k \wedge (b, c) \in k \Rightarrow (a, c) \in k$. Assume $(a, b) \in k$ and $(b, c) \in k$. Then $(\forall w)(w \in a \Rightarrow w \in b)$ and $(\forall w)$ $(w \in b \Rightarrow w \in c)$. Hence $(\forall w)[(w \in a \Rightarrow w \in b) \wedge (w \in b \Rightarrow w \in c)]$. Using the tautology $(p \Rightarrow q) \wedge (q \Rightarrow r) \Rightarrow (p \Rightarrow r)$ we deduce $(\forall w)[w \in a \Rightarrow w \in c]$. Comparing this with the definition of k we see that $(a, c) \in k$. Thus we have proved $(a, b) \in k \wedge (b, c) \in k$ $\Rightarrow (a, c) \in k$.

§3 Equivalence Relations

DEFINITION. Let A be a set. An *equivalence relation* on $A \times A$ is a relation which is reflexive, symmetric and transitive.

If A is a set of people, then the following relations are equivalence relations.

$$\{(x, y) \in A \times A | x \text{ is a blood relation of } y\}$$
$$\{(x, y) \in A \times A | x \text{ has the same parents as } y\}$$
$$\{(x, y) \in A \times A | x \text{ has the same nationality as } y\}.$$

Also $\{(x, y) \in \mathbf{R} \times \mathbf{R} | x = y\}$.
If

$$A = \left\{\frac{a}{b} | a \in \mathbf{Z} \text{ and } b \in \mathbf{Z} \setminus \{0\}\right\},$$

where $\mathbf{Z}$ is the set of all integers then we may define the equivalence relation R on $A \times A$ by

$$R = \left\{\left(\frac{a}{b}, \frac{c}{d}\right) \in A \times A | ad = bc\right\}.$$

Here, the notation $\frac{a}{b}$ does not mean "divide a by b". It denotes a formal mathematical object which consists of two integers which are written in a vertical column and separated by a bar.

In A, the elements $\frac{1}{1}, \frac{2}{2}, \frac{-1}{-1}, \ldots$ are distinct. However they are each related to the other by the relation R. We have $\left(\frac{1}{1}, \frac{2}{2}\right) \in R$ since

$1 \times 2 = 1 \times 2$ and $\left(\dfrac{2}{2}, \dfrac{-1}{-1}\right) \in R$ since $2 \times (-1) = 2 \times (-1)$. Also $\left(\dfrac{2}{12}, \dfrac{6}{36}\right) \in R$ since $2 \times 36 = 12 \times 6$. Thus R is used to make $\dfrac{1}{1}, \dfrac{2}{2}, \dfrac{-1}{-1}$ equivalent.

We observe that an equivalence relation has some at least of the essential properties of equality. Certainly a relation between objects which intuitively we may regard as defining equality between the objects must be an equivalence relation. We would require at least the usual properties of equations to hold, namely

(i) $x = x$ (reflexivity)

(ii) $(x = y) \Rightarrow (y = x)$ (symmetry)

(iii) $(x = y) \wedge (y = z) \Rightarrow (x = z)$ (transitivity).

However, an equivalence relation is required to have the property that we may substitute for an object an equivalent object before this relation is acceptable for an equality. Thus an equivalance relation R on $A \times A$ is only regarded as defining an equality between the objects of A if we have the following. For every predicate $P(x)$ which contains a variable x whose substitution set is A, and for every pair $(k, l) \in R$, we demand that $P(k) \Leftrightarrow P(l)$. For a predicate $P(x_1, x_2, x_3, \ldots, x_n)$ in which, say x_3 has substitution set A we demand that

$$(\forall x_1)(\forall x_2)(\forall x_4)(\forall x_5) \ldots (\forall x_n)(P(x_1, x_2, k, x_4, \ldots, x_n)$$
$$\Leftrightarrow (P(x_1, x_2, l, x_4, \ldots, x_n))$$

This ensures that we will not change the truth value of a statement by replacing an object by an equivalent object. We offer the following example to show that the replacement of objects by equivalent objects in a statement can indeed result in a statement with a different truth value, if we do not add the condition about substitution.

Consider "Mr. Smith knows that $\frac{1}{2}$ is the same as $\frac{2}{4}$". We may accept $0 \cdot 5$ as equivalent to $\frac{2}{4}$, but the statement "Mr. Smith knows that $\frac{1}{2}$ is the same as $0 \cdot 5$" may have a truth value different to the previous statement.

An important consequence of having an equivalence relation defined on $A \times A$ is that it cuts A up into slices, like a cake. To describe this more precisely we introduce some terminology.

Let X and Y be subsets of A. We describe X and Y as *disjoint* if $X \cap Y = \phi$. Let C be a set whose elements are subsets of A. We say that C *covers* A if each element of A occurs as an element of at least one element of C. Thus $\{\{0\}, \{1\}\}$ and $\{\{0\}, \{0, 1\}\}$ both are sets of subsets of $\{0, 1\}$ which cover $\{0, 1\}$.

Let C be a set of subsets of A such that (i) C covers A and (ii) any two elements of C are disjoint. Then we describe C as a *partition* of A. The elements of C are the slices of the cake A. We have said that (i) if we put the slices together, we get the whole cake and (ii) slices do not overlap, i.e. no crumb of the cake can be in two slices. Given an equivalence relation E on $A \times A$ we can define a partition C of A such that for all $(x, y) \in A \times A$, x is equivalent to y is and only if x and y belong to the same element of C. Thus two crumbs are equivalent if and only if they are part of the same slice. A proof follows.

Let E be an equivalence relation of $A \times A$. For each $k \in A$ we can define the set $\{y \in A | (k, y) \in E\}$ called an equivalence class of A, modulo E. This is the set of all elements of A which are equivalent to k, and we shall denote it A_k. Evidently $A_k \subseteq A$.

For each $x \in A$ we can define $A_x = \{y \in A | (x, y) \in E\}$. Then the set C defined to be $\{A_x | x \in A\}$ is a set of subsets A. It remains to show that C is a partition of A.

We observe that if $k \in A$ then we have named the slice which contains k as A_k. This naming proceedure is satisfactory in that no two different slices acquire a common name. It has the disadvantage however that there may be several names for the same slice. For example, if $l \in A_k$ then the names A_k and A_l will both refer to the same slice.

We must first show that C covers A. Let k be an arbitrary element of A. Now $A_k = \{y \in A | (k, y) \in E\}$. Since E is reflexive, $(k, k) \in E$ so that $k \in A_k$.

We must now show that any distinct pair of elements of C are disjoint. Let A_k and A_l be elements of C. We assume that A_k and A_l are not two different names for the same slice, i.e. $A_k \neq A_l$. Thus A_k and A_l are names for different slices and we must show that $A_k \cap A_l = \phi$. We prove this by contradiction. Suppose $A_k \cap A_l \neq \phi$, then there exists $m \in A_k \cap A_l$. Then $(k, m) \in E$ and $(l, m) \in E$. By symmetry we have $(k, m) \in E$ and $(m, l) \in E$. By transitivity, $(k, l) \in E$. We prove that $A_k = A_l$. Let $t \in A_k$. Then $(k, t) \in E$. By symmetry $(t, k) \in E$ which with $(k, l) \in E$ yields $(t, l) \in E$ by transitivity. Now by symmetry $(l, t) \in E$ and so $t \in A_l$. This proves that $A_k \subseteq A_l$. Next let $t \in A_l$. Then $(l, t) \in E$; since $(k, l) \in E$ transitivity yields $(k, t) \in E$ whence $t \in A_k$. Thus $A_l \subseteq A_k$ and hence $A_l = A_k$. This provides our contradiction, since we have assumed $A_k \neq A_l$.

This proves that given an equivalence relation E on $A \times A$, we can define an associated partition C of A.

Exercises 13

1. Let E be an equivalence relation on $A \times A$. For each $x \in A$, define A_x to be $\{y \in A \mid (x, y) \in E\}$. Define C to be $\{A_x \mid x \in A\}$. Prove that if k and l are elements of A then $k \in A_l \Rightarrow A_l = A_k$. Use this result to prove that two elements of C are either equal or disjoint.

2. Let C be a partition of A, $\phi \notin C$. Define R to be $\{(x, y) \in A \times A \mid$ There exists $Z \in C$ such that $x \in Z$ and $y \in Z\}$. Prove that R is an equivalence relation.

Answers 13

1. Let $k \in A_l$. To prove $A_l = A_k$. Let $x \in A_l$. Then $(l, x) \in E$. But $(l, k) \in E$ since $k \in A_l$. By symmetry and transitivity, $(k, x) \in E$. Hence $x \in A_k$. Let $x \in A_k$, then $(k, x) \in E$. But $(l, k) \in E$. By transitivity $(l, x) \in E$ so $x \in A_l$. This proves $A_l = A_k$.

   Let A_l and A_k be any two elements of C. Rather than prove

   $$(A_l \in C) \wedge (A_k \in C) \Rightarrow (A_l = A_k) \vee (A_l \cap A_k = \phi)$$

   we prove $(A_l \in C) \wedge (A_k \in C) \wedge (A_l \cap A_k \neq \phi) \Rightarrow (A_l = A_k)$. Assume $A_l \cap A_k \neq \phi$. Then there exists $m \in A$ such that $m \in A_l \cap A_k$. Thus $m \in A_l$. By the above result, $A_l = A_m$. Also $m \in A_k$ and so $A_k = A_m$. Thus $A_l = A_k$.

2. We must show that R is reflexive, symmetric and transitive.

   (i) Reflexivity. Let $k \in A$. Since C covers A, there exists $Z \in C$ such that $k \in Z$. Then evidently "$k \in Z$ and $k \in Z$" so $(k, k) \in R$.

   (ii) Symmetry. Let $(k, l) \in R$. Then there exists $Z \in C$ such that $l \in Z$ and $k \in Z$, whence $(l, k) \in R$.

   (iii) Transivity. Let $(k, l) \in R$ and $(l, m) \in R$. Then there exists $Z_0 \in C$ with $k \in Z_0$ and $l \in Z_0$. Also there exists $Z_1 \in C$ with $l \in Z_1$ and $m \in Z_1$. Thus $l \in Z_0 \cap Z_1$. Now C is a partition so its elements are either equal or disjoint. Since Z_0 and Z_1 are not disjoint, $Z_0 = Z_1$. Thence $k \in Z_0$ and $m \in Z_0$, whence $(k, m) \in R$.

   Thus to each equivalence relation on $A \times A$ there corresponds a partition of A but also to each partition of A there corresponds an equivalence relation on $A \times A$.

§4 Order Relations

A relation on $A \times A$ which is transitive is described as a *weak partial ordering* of A.

The phrase "partial ordering" indicates that we do not demand that every pair of elements of A be related. For example, if $A = \{0, 1, 2\}$ and $R = \{(1, 2)\}$ then it is true that $(\forall x)(\forall y)(\forall z)((x, y) \in R \wedge (y, z) \in R \Rightarrow (x, z) \in R)$ because $(x, y) \in R \wedge (y, z) \in R$ cannot be made true. Thus R is transitive, but 0 is not related either to 1 or to 2. We may note that ϕ, the empty set, is also a transitive relation on $A \times A$.

When dealing with a transitive relation, it is convenient to replace statements like $(k, l) \in R$ and predicates like $(x, y) \in R$ by kRl and xRy respectively. Thus if $<$ is the usual ordering of $\mathbf{R}$, we write $1 < 2$ rather than $(1, 2) \in <$ and $x < y$ rather than $(x, y) \in <$. The statement expressing transitivity then reads

$$(\forall x)(\forall y)(\forall z)(xRy \wedge yRz \Rightarrow xRz).$$

Let R be a transitive relation on $A \times A$, and let $(k, l) \in R$. Then we say "l follows k" or "l is larger than k" or alternatively "k precedes l" or "k is smaller than l".

The following definitions cope with the various ideas of largest and smallest. Let A be a set. Let R be a transitive relation on $A \times A$. Let $=$ be an equivalence relation on $A \times A$. Further, let the following substitution laws hold.

$$(\forall x)(\forall y)(\forall z)(xRy \wedge y = z \Rightarrow xRz)$$
$$(\forall x)(\forall y)(\forall z)(xRy \wedge x = z \Rightarrow zRy).$$

Thus $=$ is too have the substitution properties of equality, at least in respect of R. It helps to read xRy as "x is less than y".

(i) Let $B \subseteq A$. We call m a *maximal element of B* if $m \in B$ and for every $x \in B$, either xRm or $x = m$ or x is not related by R to m. Thus "m is a maximal element of B" means

$$(\forall x)(x \in B \wedge mRx \Rightarrow (xRm) \vee (x = m)).$$

We observe that if x is not related to m then mRx cannot be true, so the implication holds. Also, we have not claimed that xRm holds only if $x \neq m$; it is possible that xRm and $x = m$ both hold.

(ii) Let $B \subseteq A$. We call m a *minimal element of B* if

$$(\forall x)(x \in B \wedge xRm \Rightarrow (mRx) \vee (x = m)).$$

(iii) Let $B \subseteq A$. We call u an *upper bound of B* if every element of B is either less than u or equal to u. Thus "u is an upper bound B" means

$$(\forall x)(x \in B \Rightarrow xRu \vee x = u).$$

We observe that a maximal element of B is an element of B, while an upper bound of B need not be an element of B. Also, a maximal element of B need not be related to all the elements of B while an upper bound of B *must be related to every element in B.*

(iv) Let $B \subseteq A$. We call l a *lower bound of B* if

$$(\forall x)(x \in B \Rightarrow lRx \lor x = l).$$

(v) Let $B \subseteq A$. We call u a *least upper bound of B* if u is an upper bound of B and $(\forall x)(x$ is an upper bound of $B \Rightarrow uRx \lor u = x)$.

Let us suppose that B has at least one upper bound, and define a set B^{up} by

$$B^{up} = \{x \in A \mid x \text{ is an upper bound of } B\}.$$

We have ensured that $B^{up} \neq \phi$. Next, let us suppose that B has a least upper bound which we shall call u. Then $u \in B^{up}$ and u is related to all the elements of B and to all the elements of B^{up}. It follows that u is a minimal element of B^{up} and also a lower bound of B^{up}. Indeed, it is the greatest lower bound of B^{up}.

(vi) Let $B \subseteq A$. We call l a *greatest lower bound of B* if l is a lower bound of B and

$$(\forall x)(x \text{ is a lower bound of } B \Rightarrow xRl \lor l = x).$$

We observe that each definition of a largest element is paralleled by a definition of a smallest element. We are dealing with a set A which has some structure defined on it by a transitive relation R and an equivalence relation E. Now if R is transitive so is R^{-1} and if E is an equivalence so is E^{-1}, indeed $E = E^{-1}$. Thus A with the structure R^{-1} and E^{-1} is similar to A with the structure R and E. The "large" elements of the latter structure are the "small" elements of the former and vice-versa.

We may demand of R that it has more than merely transitivity. A relation R on $A \times A$ which is transitive and also satisfies the statement

$$(\forall x)(\forall y)(xRy \land yRx \Rightarrow x = y)$$

is called a *strong partial ordering* of A, or simply a *partial ordering* of A. We can show that this condition is not a consequence of transitivity but is a genuine further restriction on R by the following example. Let $A = \{0, 1), R = \{(0, 0), (0, 1), (1, 0), (1, 1)\}$ and

$E = \{(0, 0), (1, 1)\}$. Then R is transitive and E is an equivalence relation which satisfies the substitution conditions. Then $0R1 \wedge 1R0$ is true but $0 = 1$ is false.

EXAMPLES. Let A be a set of statements. Then we may define R to be $\{(x, y) \in A \times A \,|\, x \Rightarrow y\}$ and E to be $\{(x, y) \in A \times A \,|\, x \Leftrightarrow y\}$. Then R is a (strong) partial ordering of A and E is an equivalence relation on A which has the necessary substitution properties.

Let A be a set and consider $P(A)$ with the relation $R = \{(x, y) \in P(A) \times P(A) \,|\, x \subseteq y\}$. For equality on $P(A) \times P(A)$ we take set equality. Then R partially orders $P(A)$.

We now place a further restriction on R which will ensure that any pair of elements of A are related either by R or by an equality relation E. We insist that R and E be such that the following statement is true.

$$(\forall x)(\forall y)(xRy \vee yRx \vee xEy)$$

This says that x is either less than y, greater than y or equal to y. The relation R is now described as a *total ordering of A*, or a *linear ordering of A*. The usual orderings of $\mathbf{N}$, $\mathbf{Z}$ and $\mathbf{R}$ are linear. A linear ordering R of a set A is called a *well-ordering* of A if every non-empty subset of A has a first element. Thus A is *well-ordered* by R if for every subset B of A the statement

$$(\exists x)(\forall y)(x \in B \wedge y \in B \Rightarrow (xRy) \vee x = y)) \text{ holds.}$$

We recall that this property of the ordering of $\mathbf{N}$ enables us to use the induction theorem, see (ii) §3 Ch. 5. The usual ordering of $\mathbf{R}$ is not a well-ordering; indeed any set I_a of real numbers of the form $\{x \in \mathbf{R} \,|\, x < a\}$ where $a \in \mathbf{R}$ clearly has no first element since if $k \in I_a$ then $(k - 1) \in I_a$. Also any set I_a^b of real numbers of the form $\{x \in \mathbf{R} \,|\, a < x < b\}$ has no first elements since if $k \in I_a^b$ then $\dfrac{k + a}{2} \in I_a^b$ and $\dfrac{k - a}{2} < k$.

Let R be a linear ordering of A. Consider the property: Every subset of A which has an upper bound has a least upper bound.

We observe that the usual orderings of $\mathbf{N}$, $\mathbf{Z}$ and $\mathbf{R}$ all have this property. Let us denote the set of rational numbers (fractions) by $\mathbf{Q}$. The usual ordering of $\mathbf{Q}$ does not have this property. For example, let A be the set $\{x \in \mathbf{Q} \,|\, x^2 < 2\}$. Certainly this set has an upper bound; indeed 3 is an element of $\mathbf{Q}$ and is an upper bound since if $k \in A$ then where $k > 3$ we should have $k^2 > 9$ contradicting $k^2 < 2$. Hence

$x \leqslant 3$. Yet A has no least upper bound. We may prove this by contradiction. Let $b \in \mathbf{Q}$. Suppose b is a least upper bound of A. Evidently $b > 0$. Let $c = \dfrac{3b + 4}{2b + 3}$. Then c is a positive rational number. Now

$$b - c = 2\,\frac{(b^2 - 2)}{2b + 3} \quad \text{and} \quad c^2 - 2 = \frac{(b^2 - 2)}{(2b + 3)^2}.$$

(i) Suppose $b^2 < 2$. Then $c^2 - 2$ is negative and so $c < 2$. Hence $c \in A$. But if $b^2 < 2$ then $b - c$ is negative so $b < c$. Thus b is not an upper bound of A, contradiction.

(ii) Suppose $b^2 > 2$. Then $c^2 - 2$ and $b - c$ are positive. Hence $c^2 > 2$ and $b > c$. This shows that c is an upper bound of A and since $b > c$, b is not the least upper bound of A, contradiction.

(iii) Suppose $b^2 = 2$. Since b is rational it can be expressed as p/q where p and q are integers, $q \neq 0$, and further p and q have no common factor (else we should cancel). Now if $(p/q)^2 = 2$ then $p^2 = 2q^2$ and hence p^2 is divisible by 2. Hence p is divisible by 2. Thus p^2 is divisible by 4 and since $p^2 = 2q^2$, q^2 is divisible by 2. Hence q is divisible by 2. Thus p and q have the factor 2 in common, contradiction. We have proved that $b^2 < 2$, $b^2 = 2$ and $b^2 > 2$ are all false. But $<$ is a linear ordering of $\mathbf{Q}$, which means that at least one of these statements must be true. Contradiction. We conclude that b cannot be both a least upper bound of A and an element of $\mathbf{Q}$.

Thus $\mathbf{Q}$ does not satisfy the least upper bound property. This constitutes a serious defect in $\mathbf{Q}$, since the convenient exposition of much of analysis depends upon the least upper bound property being satisfied. One might well defend the thesis that the rational numbers were supplemented by the irrationals to form the real number system $\mathbf{R}$ just in order to have a system in which the least upper bound property holds. Certainly the rationals are adequate for measurement in the real world.

We may mention one further property which can be described in a structure which consists of a set A, a linear ordering R of A, and an equality relation E on $A \times A$. The set A is described as *dense* if

$$(\forall x)(\forall y)(\exists z)(xRy \Rightarrow xRz \wedge zRy).$$

Thus A is dense if between any pair of elements of A there is a third. The usual orderings of $\mathbf{N}$ and $\mathbf{Z}$ are not dense, those of $\mathbf{Q}$ and $\mathbf{R}$ are dense.

If we have a system of measurement which associates numbers with physical objects, then we should like the set of available numbers to be dense. If it is not, then we are prejudicing the accuracy of our measurements before we start.

Exercises 14

1. Let A be a set, R a transitive relation on $A \times A$ and $=$ an equality relation on $A \times A$. Take the usual identity as the meaning of $=$ and construct examples of A and R such that
    (i) There is a subset of A which has several maximal elements.
    (ii) There is a subset of A which has one maximal element but no upper bound.
    (iii) There is a subset of A which has an upper bound but no maximal element.
2. Let $\leqslant$ be a strong partial ordering of A. Prove that (i) every subset of A has *at most* one least upper bound, (ii) if $B \subseteq A$ and $b \in B$ and b is an upper bound of B then b is both a maximal element of B and the least upper bound of B; also that B has no other maximal elements.
3. Let $\leqslant$ be a strong partial ordering of A such that every non-empty subset of A which has an upper bound has a least upper bound. Prove that every non-empty subset of A which has a lower bound has a greatest lower bound.
    HINT. Let B be a non-empty subset of A which has a lower bound. Define B^{lo} to be the set of all lower bounds of B. Show that every element of B is an upper bound of B^{lo} and that B^{lo} has an upper bound. Prove that the least upper bound of B^{lo} is the greatest lower bound of B.

Answers 14

1. (i) Let $A = \{0, 1, 2\}$ and $R = \{(0, 1), (0, 2)\}$. Then 1 and 2 are both maximal elements of A.
    (ii) Let $A = N \cup \{b\}$ where b is an extra element, not an integer. We define R to be the usual ordering of N together with an extra element $(1, b)$. Thus $1Rb$ but b is related to no other element of A. It follows that b is a maximal element of A yet A has no upper bound.
    (iii) Let $A = N \cup \{b\}$, $b \notin N$, where R is the normal ordering of N together with all ordered pairs of the form (n, b) for all $n \in N$. Thus A consists of N together with the elements b tacked on as the last element of A. It follows that N is a subset of A bounded

above by b, indeed b is the least upper bound of N. Yet N has no maximal elements.

2. (i) Let $B \subseteq A$. Suppose that x and y are both least upper bounds of B. Since x is a least upper bound and y is an upper bound, $x \leqslant y$. Similarly $y \leqslant x$ whence $x = y$.

   (ii) Let b be an upper bound of B and $b \in B$. Since b is an upper bound of B then every element of B is either less than b or equal to b. Hence every element of B is either less than, equal to or unrelated to B. Hence b is a maximal element of B.

   If c is an upper bound of B then $x \in B \Rightarrow x \leqslant c$. But $b \in B$ so $b \leqslant c$. Hence b is an upper bound of B less than or equal to every other upper bound of B and so b is the least upper bound of B. Let m be a maximal element of B. Since b is an upper bound of B and $m \in B$, m and b must be related and in fact $m \leqslant b$. But $b \in B$ and m and b are related and m is a maximal element of B. Thus $b \leqslant m$. Thence $m = b$.

3. Let $B \subseteq A$ and $B \neq \phi$. Let B have a lower bound. Define B^{lo} to be the set of all lower bounds of B. Then $B^{lo} \neq \phi$. Let $m \in B$. Then for $x \in B^{lo}$, $x \leqslant m$. Thus every element of B is an upper bound of B^{lo}. Since $B \neq \phi$, B^{lo} has an upper bound. Using the hypothesis, B^{lo} has a least upper bound, say b. Now b is the *least* upper bound of B^{lo} and every element of B is *an* upper bound of B^{lo} so $b \leqslant x$ for every $x \in B$, i.e. b is a lower bound of B. Further, b is an upper bound of B^{lo}, the set of all lower bounds of B, so $x \leqslant b$ for every $x \in B^{lo}$ and so b is the greatest lower bound.

§5 Functional Relations

We shall presume that equality, to be thought of in the usual sense of identity, is defined on all the sets mentioned in this section. Let R be a binary relation on $A \times B$. We describe R as a functional relation, or simply as a function, if for every $a \in A$ there is at most one ordered pair in R having a in its first place. Thus if $(a, b) \in R$ and $(a, c) \in R$ where R is a function, then (a, b) and (a, c) must be different names for the same element of R and so $(a, b) = (a, c)$ therefore $b = c$. We may restate the definition of a function as follows. A relation $R \subseteq A \times B$ is a function if for each $a \in A$ there exists at most one $b \in B$ such that $(a, b) \in R$. If F is a set of women and M a set of men then $\{(x, y) \in F \times M \mid y$ is the natural father of $x\}$ and $\{(x, y) \in M \times F \mid y$ is the natural mother of $x\}$ are functional relations on $F \times M$ and $M \times F$ respectively.

More precisely, $R \subseteq A \times B$ is a functional relation if for all $x \in A$ and all y and $z \in B$

"if $(x, y) \in R$ and $(x, z) \in R$ then $y = z$".

Formally, $(\forall x)(\forall y)(\forall z)((x, y) \in R \wedge (x, z) \in R \Rightarrow y = z)$ where A is the substitution set of x, B of y and z, and $P(A \times B)$ of R, is a predicate in R which is true if and only if R is a functional relation. Note that $P(A \times B)$ is just the set of all relations on $A \times B$. Thus given $R \subseteq A \times B$, a proof that R is a function is a proof of the formal statement above.

EXAMPLE. Let $R \subseteq \mathbf{R} \times \mathbf{R}$ be $\{(x, y) \in \mathbf{R} \times \mathbf{R} \mid y = x + 1\}$. We prove that R is a function. Let $a \in \mathbf{R}$, $b \in \mathbf{R}$ and $c \in \mathbf{R}$. We must prove $(a, b) \in R \wedge (a, c) \in R \Rightarrow b = c$. Let $(a, b) \in R$ and $(a, c) \in R$. Then $b = a + 1$ and $c = a + 1$. Hence $b = c$. The ease of this proof demonstrates why we like such functional relations expressed in the form $\{(x, y) \in \mathbf{R} \times \mathbf{R} \mid y = (\text{an expression in } x \text{ only})\}$. We remarked in §1 that $\{(x, y) \in \mathbf{R} \times \mathbf{R} \mid 2x^2 + y^2 = 1\}$ could not be put in this form, but then it is not a functional relation.

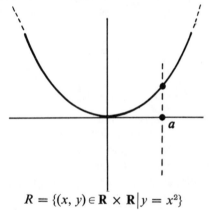

$$R = \{(x, y) \in \mathbf{R} \times \mathbf{R} \mid y = x^2\}$$

We can easily see from its picture whether a relation on $\mathbf{R} \times \mathbf{R}$ is functional. In the above picture the horizontal axis represents objects from the first place of the ordered pairs in $\mathbf{R} \times \mathbf{R}$. Given an object a in the first place of an ordered pair which is an element of the relation R, we draw a line parallel to the y axis through a. If this line cuts the picture of R then the ordered pairs in R which correspond to the points of intersection have the form $(a, \ )$.

If the vertical line through a cuts R at more than one point then R cannot be a function since the points will correspond to different ordered pairs in R which all have a in the first place. Thus R is a function while S (below) is not. We observe that for R to be a function, the vertical line through *every point* on the horizontal axis must cut the picture of R in at most one point. If even one point on the horizontal axis has a vertical line through it which cuts the picture of R in more than one place then R is not a function. In particular, if the picture of R covers an area then it cannot be a function.

We may observe in the case of the functions illustrated that S is the inverse relation of R, i.e. $S = R^{-1}$. This illustrates the fact that the inverse of a functional relation need not be a functional relation.

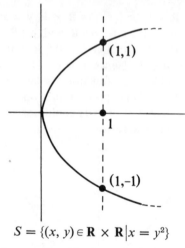

$$S = \{(x, y) \in \mathbf{R} \times \mathbf{R} \mid x = y^2\}$$

We may introduce a classification of relations in terms of whether the relation or its inverse is functional. Let R be a relation on $A \times B$. We represent A and B by circles and illustrate $(a, b) \in R$ by labelling a point in A's circle a, a point in B's circle b, and joining the points with an arrow. This enables us to draw pictures illustrating our classification.

1.

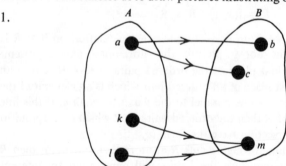

The relation R is such that at least one element of A (e.g. a) is related to more than one element of B *and* at least point of B (e.g. m) has at least two elements of A which are related to it. In short, neither R nor R^{-1} is a function. Such a relation is described as a *many-to-many* relation.

2.

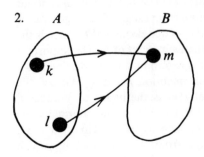

The relation R is such that no element of A is related to more than one element of B, but at least one element of B has more than one element of A related to it. In short, R is a function and R^{-1} is not. Such a relation is called *many-to-one*.

3.

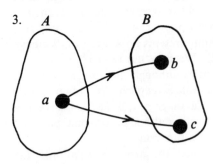

The relation R is not a function, but R^{-1} is a function. The relation is described as *one-to-many*.

4.

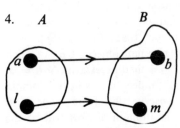

Both R and R^{-1} are functions. Such a relation is described as *one-to-one* and also called an *injection*.

The following definitions are framed for arbitrary relations. Let $R \subseteq A \times B$. The *domain* of R is the set $\{x \in A \,|\, (x, y) \in R\}$. The *range* of R is the set $\{y \in B \,|\, (x, y) \in R\}$. A *co-domain* of R is a set which has the range of R as a subset. Thus the relation $\{(x, y) \in \mathbf{R} \times \mathbf{R} \,|\, y = x^2\}$ has domain $\mathbf{R}$, and $\mathbf{R}_0^+$ is the range of this relation. It is not usual to specify functions as sets of ordered pairs. Rather, it is customary to present the domain of the function together with a rule which assigns to each object in the domain a unique second object. We call the second object the *image* of the object in the domain *under* the function.

We use the notation $f: A \to B$ to indicate that f is a function whose domain is A and that B is a co-domain of f. This notation is fussy about the domain of f but not so fussy about the range of f. In general, if $f: A \to B$ and $x \in A$ then we denote the unique element of B which is the image of x under f by $f(x)$, or alternatively by xf of f_x. The important point is that the image of x under f is determined by x and f and so a suitable name for the image should mention f and x. We use the notation $f: x \mapsto f(x)$ to indicate that $f(x)$ is the image of x under f. Thus the function $\{(x, y) \in \mathbf{R} \times \mathbf{R} \,|\, y = x^2\}$ would normally be described by two pieces of information, namely $f: \mathbf{R} \to \mathbf{R}$ and $f: x \mapsto x^2$. The range of f may be determined from this information, which explains the casual use of a co-domain rather than the range itself. Had we described f as $f: \mathbf{R} \to \mathbf{R}_0^+$ and $f: x \mapsto x^2$ then the same function results. Translation from the description $f: A \to B$ and $f: x \mapsto f(x)$ back to the ordered pair description presents no difficulties. We have simply $\{(x, y) \in A \times B \,|\, y = f(x)\}$. We may use $(x \mapsto x^2)$ as a more informative name for the function f; although the domain does not appear in this name at least the rule is explicit. We observe that $f(x)$ is the name of an element in the range of f, it may not serve as a name for f. Thus "x^2" is not a function although for some purposes it acts as a sufficient description of a function.

The following definition is framed for arbitrary relations. Let $R \subseteq A \times B$ and $S \subseteq C \times D$ be relations. Then $S_0 R$ denotes the *composition* of R and S, which is a relation on $A \times D$ defined to be $\{(x, y) \in A \times D \,|\, \text{There exists } z \in B \cap C \text{ such that } xRz \text{ and } zSy\}$. Thus x is related to y by SoR if there is an intermediate object $z \in B \cap C$ such that xRz and zSy. Of course if $B \cap C = \phi$ then necessarily $SoR = \phi$. If S and R are functions then so is SoR; indeed if x is in the domain of SoR then $(SoR)(x) = S(R(x))$. We observe that if the range of R is a subset of the domain of S then the domain of SoR will be the domain of R while a suitable co-domain for SoR will be that of S. Thus if $f: A \to B$ and $g: B \to C$ then $gof: A \to C$. Further, if $f: x \mapsto f(x)$ and $g: x \mapsto g(x)$ then $gof: x \mapsto g(f(x))$.

Since relations are sets, they are equal if they are equal as sets— set equality having been defined. We observe that if two relations, and in particular two functions, are equal then their domain and range are equal.

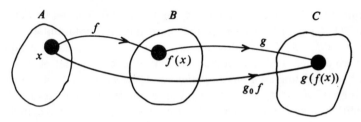

A function f is a *surjection onto a set B* if B is the range of f. If f has been described via the form $f : A \to B$ then we may describe f as surjective in order to indicate that B is actually the range of f. Alternatively, given A and B we may wish to restrict ourselves to functions whose domain is A and whose range is B; such functions are usually described as surjections from A onto B. Thus surjectivity is not a property of functions but a means of indicating the range of a function. In contrast, *injectivity* is a property which a function may or may not have. Notwithstanding this, it is usual to describe a function f: $A \to B$ as a *bijection* in order to indicate that the function is an injection whose range is B.

Let $A : I \to C$ be a surjection, where C is a set whose elements are sets. If $i \in I$, then we denote the image of i under A by A_i. We describe C as a family of sets indexed by I, and refer to I as the index set of C. The process is usually accomplished by "Let $(A_i : i \in I)$ be a family of sets". We may use the index idea to extend the meaning of union, intersection and cartesian product. Thus the union of the family $(A_i : i \in I)$ is $\{x \mid \text{There exists } i \in I \text{ such that } x \in A_i\}$, denoted $\cup(i \in I)A_i$; and the intersection of this family is $\{x \mid \text{for all } i \in I, x \in A_i\}$, denoted $\cap(i \in I)A_i$. In order to extend the notion of cartesian product we first re-interpret the original definition of $X \times Y$. Let $(A_i : i \in \{1, 2\})$ be a family of two sets. We define the cartesian product of this family to be $\{f \mid f : \{1, 2\} \to A_1 \cup A_2 \text{ such that } f(1) \in A_1 \text{ and } f(2) \in A_2\}$. The elements of this set are thus functions and a typical element has the form $\{(1, f(1)), (2, f(2))\}$ where $f(1) \in A_1$ and $f(2) \in A_2$. We have only to associate $\{(1, f(1)), (2, f(2))\}$ with the ordered pair $(f(1), f(2))$ to see that these two definitions of $A_1 \times A_2$ give rise to essentially similar objects. The new definitions may however be extended. We may define the cartesian product $X(i \in I)A_i$ of the family $(A_i : i \in I)$ to be

$$\{f \mid f : I \to U(i \in I)A_i \text{ and for each } i \in I, f(i) \in A_i\}.$$

For our present purposes it will suffice to consider the cartesian product of a finite family $A_1, A_2, A_3, \ldots, A_n$ of sets which we may define to be $\{(a_1, a_2, \ldots, a_n) \mid a_i \in A_i \text{ for } 1 \leqslant i \leqslant n\}$. This set we denote by $\mathbf{X}(1 \leqslant i \leqslant n)A_i$ or else use the more intuitive notation $A_1 \times A_2 \times A_3 \times \ldots \times A_n$. We describe $(a_1, a_2, \ldots, a_n)$ as ordered n-tuple. As we defined a binary relation on $A \times B$ to be a subset of $A \times B$, just so we define an *n-place relation on* $A_1 \times A_2 \times \ldots \times A_n$ to be a subset of $A_1 \times A_2 \times \ldots \times A_n$. For example, we may regard the addition of real numbers as a 3 place relation. Thus $+$ is a subset of $\mathbf{R} \times \mathbf{R} \times \mathbf{R}$ such that $(a, b, c) \in +$ if and only if $a + b = c$. We may also regard $+$ as a function whose domain is $\mathbf{R} \times \mathbf{R}$ and whose range is $\mathbf{R}$. Thus $+ : \mathbf{R} \times \mathbf{R} \to \mathbf{R}$ and using functional notation we have, for example, $+((4, 3)) = 7$. Of course we are not in the habit of writing the image of $(a, b) \in \mathbf{R} \times \mathbf{R}$ under $+$ as $+((a, b))$; we prefer to write it as $a + b$.

Exercises 15

1. Let R and S be functions. Prove that $R \cap S$ is then a function and give a counter-example to show that $R \cup S$ need not be a function.

2. Find the domain and range of $\left\{(x, y) \in \mathbf{R} \times \mathbf{R} \left| \dfrac{x^2}{9} + \dfrac{y^2}{16} = 25\right.\right\}$, and determine whether this relation, or its inverse, is a function.

3. Let $R : A \to B$ be an injection. What is the range and domain of $R^{-1}oR$. Describe $R^{-1}oR$ and also RoR^{-1}.

4. Let $f : A \to B$ and suppose $(A \times B) \setminus f$ is a function. Let $f \neq \phi$ and $(A \times B) \setminus f \neq \phi$. Prove that the set B contains exactly two elements.

5. Let $R \subseteq A \times B$, $S \subseteq C \times D$ and $T \subseteq E \times F$. Then SoR and ToS are relations. Of which cartesian products are SoR and RoS subsets? Of which cartesian products are $To(SoR)$ and $(ToS)oR$ subsets? Show that the relations $(ToS)oR$ and $To(SoR)$ are equal.

6. Let $f : A \to B$ and $g : A \to B$. We define the function $(f + g)$ by (i) $(f + g) : A \to B$ and (ii) $(f + g) : x \mapsto f(x) + g(x)$. Explain what it is that we have defined and also what assumptions must be made in order that the definition be meaningful.

Answers 15

1. If (a, b) and $(a, c) \in R \cap S$ then (a, b), $(a, c) \in R$ and since R is a function $b = c$. If $R = \{(1, 1)\}$ and $S = \{(1, 2)\}$ then $R \cup S$ is not a function.

2. Domain is $\{r \in \mathbf{R} \mid -15 \leqslant r \leqslant 15\}$. Range is $\{r \in \mathbf{R} \mid -20 \leqslant r \leqslant 20\}$. The relation contains $(0, 20)$ and $(0, -20)$ and its inverse contains $(0, 15)$ and $(0, -15)$ hence neither is a function.

3. $R^{-1}oR: A \to A$ is surjective such that $(R^{-1}oR): x \mapsto x$. RoR^{-1}: $B' \to B'$ where $B' \subseteq B$ is the range of R; $RoR^{-1}: x \mapsto x$.

4. Since $f \neq \phi$, neither A or B is empty. Let $a \in A$ and $b \in B$ and $(a, b) \in f$. Now B cannot contain just the element b or $(A \times B) \setminus f$ would be empty. If B contains two elements c and d in addition to b then (a, c) and (a, d) are elements of $(A \times B) \setminus f$ which is hence not a function.

5. $SoR \subseteq A \times D$, $ToS \subseteq C \times F$, $To(SoR) \subseteq A \times F$. Let $(a, b) \in To(SoR)$. Then there exists $z_1 \in D \cap E$ such that $a(SoR)z_1$ and z_1Tb. Since $(a, z_1) \in SoR$ there exists $z_2 \in B \cap C$ such that aRz_2 and z_2Sz_1. Since z_2Sz_1 and z_1Tb, $(z_2, b) \in (ToS)$ i.e. $z_2(ToS)b$. Since aRz_2 we have $(a, b) \in (ToS)oR$ as required. The converse is similar.

6. We have defined addition on the set F of all functions from A to B. We have assumed that an addition is defined on the set B, so that $f(x) + g(x)$ is meaningful. Note that the symbol $+$ has two meanings, one as addition on F and the other as addition on B. It is assumed that the context makes it clear which meaning is to be given to each occurrence of $+$.

CHAPTER 7

THE
AXIOMATIC APPROACH

§1 The Language of Quantification

We reintroduce more formally the language of Quantification which was the subject of Chapter 5. The vocabulary of this Language consists of the connectives, $\sim$, $\wedge$, $\vee$ and $\Rightarrow$; the quantifiers $\forall$ and $\exists$; brackets and commas for punctuation; an inexhaustible list of symbols $x, y, z, \ldots$ for use as variables; a similar list $a, b, c, \ldots$ for use as parameters and yet another list $P, Q, R, \ldots$ for use as predicates.

We may specify how sentences are formed using this vocabulary as follows. A term consists of a predicate symbol followed by a list of variables and parameters. Thus $P(x)$ is a term in which P is a one-place predicate; $Q(x, a)$ is a term in which Q is a two place predicate, one of whose places is marked by the variable x and the other filled by the parameter a. We may specify how sentences are constructed by the following rules.

(i) If t is a term, then t is a sentence.
(ii) If S_1 and S_2 are sentences then so are $(\sim S_1)$, $(S_1 \wedge S_2)$, $(S_1 \vee S_2)$, $(S_1 \Rightarrow S_2)$.
(iii) If S is a sentence, x a variable, then $(\forall x)(S)$ and $(\exists x)(S)$ are sentences.

We stipulate that if S is a sentence in which the variable x does not occur then $(\forall x)(S)$ and $(\exists x)(S)$ are the same as S. Thus $(\forall y)(Q(x, a))$ is the same as $Q(x, a)$.

A sentence S, all of whose variables are quantified, is a statement — see §1 Chapter 5.

Let us examine the statement

$$(\forall x)(\forall y)(\forall z)(P(x, y, z) \Rightarrow P(y, x, z)).$$

We presume this statement is supposed to be about some objects but it does not mention any objects. We presume that the predicate $P(x, y, z)$ says something about triples of objects but we do not know what it says. The statement does however expose a characteristic of $P(x, y, z)$.

A *model* of this statement consists of a non-empty set U of objects and a relation $R \subseteq U \times U \times U$ such that when the statement is interpreted as a statement about the objects in U and the relation R then the statement is true. For example, if we take $\mathbf{R}$ as our set of objects and interpret $P(x, y, z)$ as meaning $x + y = z$ then the statement means for all real numbers x, y and z, if $x + y = z$ then $y + x = z$, i.e. if x and y are real numbers then $x + y = y + x$.

Another model of this statement consists of $\mathbf{R}$ and the relation $\{(x, y, z) \in \mathbf{N} \times \mathbf{N} \times \mathbf{N} \,|\, xy = z\}$ i.e. multiplication on $\mathbf{N}$. The statement then says that for any two natural numbers x and y, $xy = yx$.

An *interpretation* of the statement consists of a non-empty set U and a relation $R \subseteq U \times U \times U$. The question then arises of whether the statement is true under the given interpretation (U, R). If the statement is true under this interpretation (U, R) then (U, R) is a model of the statement.

Thus one interpretation of the above statement consists of taking $\mathbf{N}$ as the set of objects and $x^y = z$ as the meaning of $P(x, y, z)$. This interpretation is not a model of the statement since the statement is not true under this interpretation—for example $2^3 = 8 \neq 9 = 3^2$.

In general, an interpretation of a statement in the language of quantification consists of a set U of objects together with a list of relations, one relation for each of the predicates which occur in the statement. Of course the relation R which is the interpretation of a predicate P must be suitable in the sense that if $P(x_1, \ldots x_n)$ is a predicate with n variables then R must be a set of n-tuples. Thus the interpretation of $P(x, y)$ must be a subset of $U \times U$, that of $P(x, y, z)$ a subset of $U \times U \times U$ and so on.

We have seen that the truth of a statement on this language depends upon how it is interpreted. A statement of this language is described as *valid* if it is true under all interpretations. For example, $(\forall x)P(x) \Leftrightarrow \sim(\exists x)(\sim P(x))$ and $(\forall x)(P(x) \wedge Q(x)) \Leftrightarrow (\forall x)P(x) \wedge (\forall x)Q(x)$ are both valid. The valid statements are those which are true of logical necessity. They are the statements of the language which are "always true". The valid statements of quantification are the counterparts of the tautologies of the language of statements.

It is possible to set up a formal proof procedure under which all the valid statements, and only the valid statements, of this language are provable. To do so would not be within the scope of this book. The interested reader might consult R. M. Smyllyan, *First Order Logic*, Springer 1968. We shall however connect the proof of valid statements with §2 of Chapter 5 by means of an example.

To prove that $(\forall x)P(x) \Rightarrow \sim(\exists x)(\sim P(x))$ is valid, let U be a non-empty set and let $R \subseteq U$ be the one-place relation which is the interpretation of $P(x)$. The usual method of proof of valid statements is by contradiction. Suppose that $(\forall x)P(x) \Rightarrow \sim(\exists x)(\sim P(x))$ is false under the interpretation (U, R). Then $(\forall x)P(x)$ is true and $\sim(\exists x)(\sim P(x))$ is false. Hence $(\exists x)(\sim P(x))$ is true and so there exists $k \in U$ such that $\sim P(k)$ is true, and so $P(k)$ is false. However, $(\forall x)P(x)$ is true and $k \in U$ so in particular $P(k)$ is true. Thus we have our contradiction. We have proved the statement true under an arbitrary interpretation and so we conclude that it is true under all interpretations and is therefore valid.

Let us consider a simple axiom system.

s_1 $(\forall x)E(x, x)$

s_2 $(\forall x)(\forall y)[E(x, y) \Rightarrow E(y, x)]$

s_3 $(\forall x)(\forall y)(\forall z)[E(x, y) \wedge E(y, z) \Rightarrow E(x, z)]$

s_4 $(\forall x)(\forall y)(\forall z)(\forall u)(\forall v)(\forall w)[E(x, u) \wedge E(y, v) \wedge E(z, w)$
$\wedge S(x, y, z) \Rightarrow S(u, v, w)]$

s_5 $(\forall x)(\forall y)(\exists z)S(x, y, z)$

s_6 $(\forall x)(\forall y)(\forall z)(\forall w)[S(x, y, z) \wedge S(x, y, w) \Rightarrow E(z, w)]$

s_7 $(\forall x)(\forall y)(\forall z)(\forall u)(\forall v)(\forall w)[S(x, y, z) \wedge S(z, u, v) \wedge S(y, u, w)$
$\Rightarrow S(x, w, v)]$

s_8 $(\exists x)(\forall y)(\exists z)[S(x, y, y) \wedge S(z, y, x)]$.

The above axioms are written in the language of quantification. If we anticipate the intended interpretation of these axioms by writing $x = y$ in place of $E(x, y)$ and $x + y = z$ in place of $S(x, y, z)$ the result looks much more natural.

s_1' For all x, $x = x$.

s_2' For all x and y, $x = y$ implies $y = x$.

s_3' For all x, y and z, if $x = y$ and $y = z$ then $x = z$.

s_4' For all x, y, z, u, v and w, if $x = u$, $y = v$ and $z = w$ and $x + y = z$ then $u + v = w$.

s_5' For all x and y there exists z such that $x + y = z$.

s_6' For all x, y, z and w, if $x + y = z$ and $x + y = w$ then $z = w$.

s_7' For all x, y and u, $(x + y) + u = x + (y + u)$.

s_8' There exists an x such that for all y, $x + y = y$ and such that for all y there exists z such that $z + y = x$.

We observe that some care is required in order to render s_7' in the required language and obtain s_7.

We may consider $s_1 - s_8$ as describing a sort of structure. It makes no difference whether we consider the structure described by the eight sentences $s_1 - s_8$ or by the single sentence $s = s_1 \wedge s_2 \wedge \ldots \wedge s_8$. It would make a great deal of difference if the description consisted of a collection of infinitely many sentences since we cannot, in this language, take the conjunction of infinitely many sentences. We may now look for models of s. That is, for a set U and relations $= \subseteq U \times U$ and $R \subseteq U \times U \times U$ corresponding to E and S such that s is true. We may take $\mathbf{R}$ with the usual equality and $x + y = z$ as the meaning of $S(x, y, z)$ and this will constitute a model of s. The x whose existence is required by s_8' is zero and we require $0 + y = y$ for every y and also that for every $y \in \mathbf{R}$ there exists an appropriate $z \in \mathbf{R}$ such that $z + y = 0$. An appropriate model can also be obtained using $\mathbf{Z}$ with the same addition or $\mathbf{C}$, the complex numbers, with their addition. Thus not only does $\mathbf{R}$ provide a model but $\mathbf{Z}$, a subset of $\mathbf{R}$ and $\mathbf{C}$, an extension of $\mathbf{R}$, also provide models of s. If we take $\mathbf{R} \setminus 0$ and interpret $S(x, y, z)$ as $xy = z$ we also obtain a model; the x whose existence is required by s_8' is 1. We may also obtain finite models, for example by taking $\{0, 1, 2\}$ and interpreting $S(x, y, z)$ as addition modulo 3.

A structure which is a model of s is called a *group*. Given a particular group one might investigate what substructures exist in the group, as $\mathbf{Z}$ exists within $\mathbf{R}$, or how one might extend the group to obtain a larger structure. One might also attempt to explain the behaviour of a structure in terms of component substructures.

We have observed that there are models of s, which, although they satisfy the description afforded by s, are nevertheless very different in other respects. One might therefore consider adding axioms to s with which to distinguish between different types of group. On the other hand, given a structure, say that normally associated with $\mathbf{R}$, one may attempt to find a complete description of $\mathbf{R}$. This amounts to finding a list of axioms of which $\mathbf{R}$, with the usual addition, multiplication and order, is a model such that any model of these axioms is essentially the same as $\mathbf{R}$. Then indeed we would be able to claim that we knew all the basic properties of the real number system, from which all other properties could be deduced.

One may investigate groups by studying the group axioms without referring to any particular group. If we can deduce a statement from the group axioms, then this statement must hold for every group. Thus we may comfortably expose general properties of groups. On the other hand, a statement which is true of some groups but false of others will not be deducible from the group axioms. It is then of interest to discover exactly what additional properties are required of a group in order that such a statement be true.

Let t be a statement which is deducible from the group axiom s. The deduction of t from s as hypothesis can be thought of as a direct proof that $s \Rightarrow t$ is a valid statement. We emphasise that all such proofs prove that a certain statement is always true. The words "hypothesis" and "assumption" are used to indicate the left hand side of an implication. Thus having described which statements of a language are always true, it is sufficient to specify a formal system of deduction which renders just the "always true" statements provable.

Thus far we have not mentioned how parameters are employed. They fulfill essentially two roles. Firstly, they are used in the proof of valid statements. In our example of the proof of $(\forall x)(P(x) \Rightarrow \sim(\exists x)(\sim P(x))$ we took an arbitrary interpretation (U, R) of this sentence. It is quite unnecessary to involve ourselves with an interpretation however; no use was made of R in the proof and only perfunctory use was made of U. In fact we only invoked the set of objects U so that there would be an object of which k could be the name. Formally, we can work with names without requiring the existance of objects which have these names. The parameters then are a list of names waiting for an interpretation to supply objects to be called by these names. Then from $(\exists x)(\sim P(x))$ we deduce $\sim P(k)$ where k is a parameter. In general if an existential statement $(\exists x)Q(x)$ occurs in a proof we allow ourselves to deduce $Q(k)$ where k is a parameter which has not so far occurred in the proof. The provision that k is new is necessary since when the sentence *is* given an interpretation and objects become available $(\exists x)Q(x)$ will assert the existence of a certain object. What we have done in deducing $Q(k)$ is to reserve k as the name of that certain object. It would not do if k had already been reserved as the name of a different object. In contrast, from a universal statement $(\forall x)Q(x)$ we may deduce any or all of $Q(a)$, $Q(b)$, $Q(c)$, ... and so on for all the parameters a, b, c, ... of the language. It does not matter which objects acquire the names a, b, c, ... under the interpretation since the truth of $(\forall x)Q(x)$ necessitates that $Q(x)$ is true whatever object fills the space marked x.

Parameters may also be used in an axiom system if their use is convenient. Thus the axiom s_8 may be replaced by

$$s_8'' \qquad (\forall y(\exists z[S(a, y, y) \wedge S(z, y, a)].$$

We have simply replaced s_8 which asserts the existence of a certain object by s_8'' which asserts this existence implicitly by using the name a. When we supply an interpretation for $s_1, \ldots, s_7$, and s_8'' we must specify not only a set of objects U and two relations to correspond to E and S but also we must specify an object in U which is to receive the name a before we can discuss whether s_8'' is true or not.

Although the language introduced in this section is sufficient to describe many important structures, it is not sufficient to describe even the real number system. The advantage knowing that a structure can be described in this language is that general theorems exist which apply to all structures which can be described in this language. We observe however that although a description is allowed to contain infinitely many statements $s_1, s_2, s_3, \ldots$; the conjunction of infinitely many statements $s_1 \wedge s_2 \wedge s_3 \wedge \ldots$ is not a statement of this language and neither is the disjunction $s_1 \vee s_2 \vee s_3 \vee \ldots$. Also, the quantifiers may only quantify the variables $x, y, z, \ldots$ they may not quantify the predicates $P, Q, R, \ldots$. Thus we are not allowed to speak of all the relations of such and such a kind, nor can we inquire whether there exists a relation of a certain type.

§2 The Language of Sets

The language which we now describe suffices to express most of mathematics. Its vocabulary consists of $\sim$, $\wedge$, $\vee$, $\Rightarrow$, $\forall$, $\exists$ and brackets and commas for punctuation together with a list $x, y, z \ldots$ of variables, a list $a, b, c, \ldots$ of parameters and one binary predicate, $\in$. There are no other predicate symbols within the language. The sentences are composed just like those of the language of quantifications except that we need only consider terms which consist of $\in$ followed by variables or parameters. A term like $\in(a, b)$ or $\in(x, a)$ is written in the form $a \in b$, $x \in a$ respectively.

An interpretation of a sentence in this language will consist of a collection U of objects and a binary relation to correspond to $\in$. The objects of U we formally describe as sets. Notice that we have studiously avoided describing U as a set in the present context. Thus our language is intended to talk about sets, it does not mention individuals which are not sets. There is no loss involved in this since most mathematical objects can conveniently be defined so that they are sets.

We should like a valid statement of the language of sets to be one which is true under all interpretations of $\in$. Let us enquire whether $(\forall x)(x \in x)$ is valid. Consider an interpretation which consists of a collection U which contains a set B. We can define the collection A to be $\{x \in B | x \notin x\}$. Let us suppose our interpretation is such that A is contained in the collection U, so that A is also a set. We now ask whether $A \in A$ is a true or false statement. If it is false then $(\forall x)(x \in x)$ clearly cannot be valid. Now if $A \in A$ is true then by definition of A, $A \notin A$ and we have a contradiction. Under our system of deduction as outlined in Chapter 3 and §2 of Chapter 5, we may deduce that $A \in A$ is false. Unhappily, if $A \in A$ is false, i.e. $A \notin A$, then by definition of A, $A \in A$ is true, another contradiction. Our system of deduction allows us to infer that $A \notin A$ is true. Thus this interpretation yields a statement $A \in A$ which is both true and false.

We do not think that we ought to be able to prove that a statement is both true and false, indeed our natural method of argument is based on the assumption that this is not possible. There are two conceivable possibilities. Either our natural methods of deduction are faulty and we may hope to replace them by something better, or else we have been over generous and allowed the statements of the language to be interpreted too liberally. Taking the latter point of view we can curtail the interpretations which a statement is allowed by specifying a special collection of sentences, which we shall call the axioms of set theory, and insisting that we consider only interpretations under which the axioms of set theory are true. We will call such interpretations admissible. A valid sentence of set theory is now a statement which is true under all *admissible* interpretations. Thus the axioms of set theory are valid by definition. We shall not discuss which sentences might constitute a suitable collection of axioms. The interested reader is refered to A. Abian, *The Theory of Sets and Transfinite Arithmetic*, W. B. Saunders Company, 1965. It is not currently known whether the axioms detailed in this book do in fact result in our not being able to prove a statement both true and false but they are widely accepted as promising.

We consider the natural question of how the language of a set theory, with its single predicate, is more expressive than that of quantification theory. In fact we can produce any number of predicates in set theory using the single predicate $\in$. Whereas in quantification we might invoke a predicate $P(x)$, in set theory we invoke a set A. We can acquire the set corresponding to $P(x)$ by defining A to be $\{x | P(x)\}$ but we shall have left the language of quantification theory. On the other hand, the predicate associated with A is just "$x \in A$" and we are still within the language of set theory. We are allowed, in set theory, to quantify sets and speak of the existence of certain sets, the set of all subsets of a set, the set of all functions from A to B and so on, whereas we are not allowed to quantify predicates in quantification theory. The notions of equality, subset, union, intersection, power set, ordered pair and cartesian product are all definable in terms of $\in$. In the case of equality, defined in §4 Chapter 5, we observe that the properties ascribed to equality by s_1, s_2 and s_3 of the last section follow from the definition. The substitution property of equality need only be ascribed for the predicate $\in$ since all other predicates are defined in terms of $\in$ and this is done by one of the axioms of set theory.

The axioms $s_5 - s_8$ may be expressed in set theory using the derived notions of equality, subset and so on as follows.

st_5 $R \subseteq A \times A \times A$

st_8 $(\forall x)(\forall y)(\forall z)(\forall w)[(x, y, z) \in R \wedge (x, y, w) \in R \Rightarrow z = w]$

st_7 $(\forall x)(\forall y)(\forall z)(\forall u)(\forall v)(\forall w)[(x, y, z) \in R \wedge (z, u, v) \in R \wedge$
$\quad (v, u, w) \in R \Rightarrow (x, w, v) \in R]$

st_8 $(\exists x)(\forall y)(\exists z)[(x, y, y) \in R \wedge (z, y, x) \in R].$

The symbols R and A used above are parameters. We could dispense with the use of parameters, rendering st_5 for example as $(\exists x)(\exists y)$ $(x \subseteq y \times y \times y)$. A model of these sentences is an interpretation in which the axioms of set theory and these sentences are all true.

Index

Antisymmetric relation; 90

Bijection; 109
Binary
  ∼ connective; 10
  ∼ predicate; 47

Cartesian product; 47
Co-domain; 107
Complement; 44
Composition of functions; 108
Compound statement; 9
Conjuction; 12
Connective; 10
  unary ∼; 10
  binary ∼; 10
  joint denial ∼; 21
  nor ∼; 21
  Sheffer's ∼; 21
  nand ∼; 21
Contradiction, proof by; 15, 30
Counter example; 60
Cover; 96

Dense; 102
Difference of two sets; 45
Direct proof; 14, 27
Disjoint sets; 96
Disjunction; 13
Domain; 107

Empty set; 45
Equal sets; 71
Equivalence relation; 95
Equivalence statements; 18, 32
Existential quantifier; 52

Function; 104
  injective ∼; 107

composition of two ∼; 108
  surjective ∼; 109
  bijective ∼; 109

Greatest lower bound; 100
Group; 115

Image; 108
Implication; 14
Inclusion; 72
Indirect proof; 15, 28
Injection; 107
Interpretation; 113
Intersection; 45
Inverse relation; 80

Joint denial; 21

Least upper bound; 100
Linear ordering; 101
Logical constant; 23
Lower bound; 100

Many-to-many relation; 107
Many-to-one relation; 107
Maximal element; 99
Minimum element; 99
Model; 113
Modus ponens; 24

Nand connective; 21
Negation; 12
Nor connective; 21
n-place relation; 110

One-to-one relation; 107
One-to-many relation; 107
Ordered pair; 47
Ordering
  ∼ relation; 98

Ordering—(*contd.*)
  linear ∼; 101
  total ∼; 101
  weak partial ∼; 100
  partial ∼; 100

Parameter; 38
Partial ordering; 100
Partition; 97
Power set; 76
Predicate; 41
  unary ∼; 41
  binary ∼; 47
Proof
  direct ∼; 4, 27
  indirect ∼; 15, 28
  ∼ by contradiction; 15, 30
  constructive/non-constructive
    ∼; 59
Proper subset; 73

Range; 107
Reflexive relation; 87
Relation; 77
  inverse ∼; 80
  reflexive ∼; 87
  symmetric ∼; 89
  transitive ∼; 89
  antisymmetric ; 90
  equivalence ∼; 95
  functional ∼; 104
  many-to-one ∼; 107
  many-to-many ∼; 107
  one-to-many ∼; 107
  one-to-one ∼; 107
  domain of a ∼; 107
  range of a ∼; 107
  co-domain of a ∼; 107
  n-place ∼; 110
  order ∼; 98
Rule of inference; 24

Set; 40, 117
  element of a ∼; 41
  member of a ∼; 41
  union of two ∼; 43
  complement of a ∼; 44

intersection of two ∼; 45
empty ∼; 45
difference of two ∼; 45
Cartesian product of two ∼; 47
equal ∼; 71
∼ inclusion; 72
sub ∼; 73
proper sub ∼; 73
super ∼; 73
power ∼; 76
disjoint ∼; 96
partition of a ∼; 97
substitution ∼; 37
Sheffer's connective; 21
Statement; 9, 11
  elementary ∼; 9
  compound ∼; 9
  equivalent ∼; 18, 32
  ∼ variable; 23
Strong partial ordering; 100
Subset; 73
Substitution set; 37
Superset; 73
Surjection; 109
Symmetric relation; 89

Tautology; 18
Theorem; 25
Total ordering; 101
Transitive relation; 89
Truth value, ∼ table; 12

Unary
  ∼ connective; 10
  ∼ predicate; 41
Union; 43
Universal
  ∼ quantifier; 52
  ∼ set; 43
Upper bound; 99

Valid; 113
Variable; 23, 37
Venn diagram; 44

Weak partial ordering; 98